AF730633

Thought

Thought

How we recognize, remember, anticipate, plan, perform, communicate, and experience consciousness

By Bruce Maier

Copyright © 2020 Bruce Maier

All rights reserved.

March 2020

ISBN: 978-0-578-64345-8

The cover picture is from
"Atlas and Text-book of Human Anatomy Volume III Vascular System, Lymphatic system, Nervous System and Sense Organs," by Dr. Johannes Sobotta, 1908, now in the Public Domain.

Preface

While I spent my career in computer technology, I always had a keen interest in neurobiology. At first, I held out hope that Deep Learning would provide useful insight into cognitive processes. Instead, the Computational Neural Networks of Deep Learning (an introduction is in the Addendum) proved to be more artificial intelligence than real intelligence.

After retirement in 2012, I began a deep dive into neurobiology with an emphasis on the mechanisms of cognitive processing. While functional-MRI imaging has given us insight into how multiple regions of the brain are involved, neuron-level understandings are missing.

I thus began an effort to construct a neuron-level model of cognitive functioning. I thought about how I arrived at conclusions, or navigated my way, or accomplished a task. From that, and some basic neurobiology comes the main subject of this book, the Connections Model.

Acknowledgments

I wish to thank all of the reviewers for their contributions. In particular, my friends Carl, Harvey, Gary, and my wife, Sandra. Sandra was most patient with me as I shared new learning and insight on my path to writing this book.

Introduction

The book presents a new theory, the Connections Model. The Connections Model explains how we recognize, remember, anticipate, plan, respond, perform, communicate, and experience consciousness.

Before delving into the Connections Model, an introduction to basic neurobiology follows. Following the Connections Model is an addendum of related topics.

Finally, my references include Wikipedia, Scientific American, and the New York Times. These are not research paper level references.

But this book is not a research paper. The general public is the audience. The general public does not have access to most research paper level references.

Note that Wikipedia articles contain footnotes with primary references that should satisfy the demanding reader.

Table of Contents

Neurobiological Concepts

The Human Brain

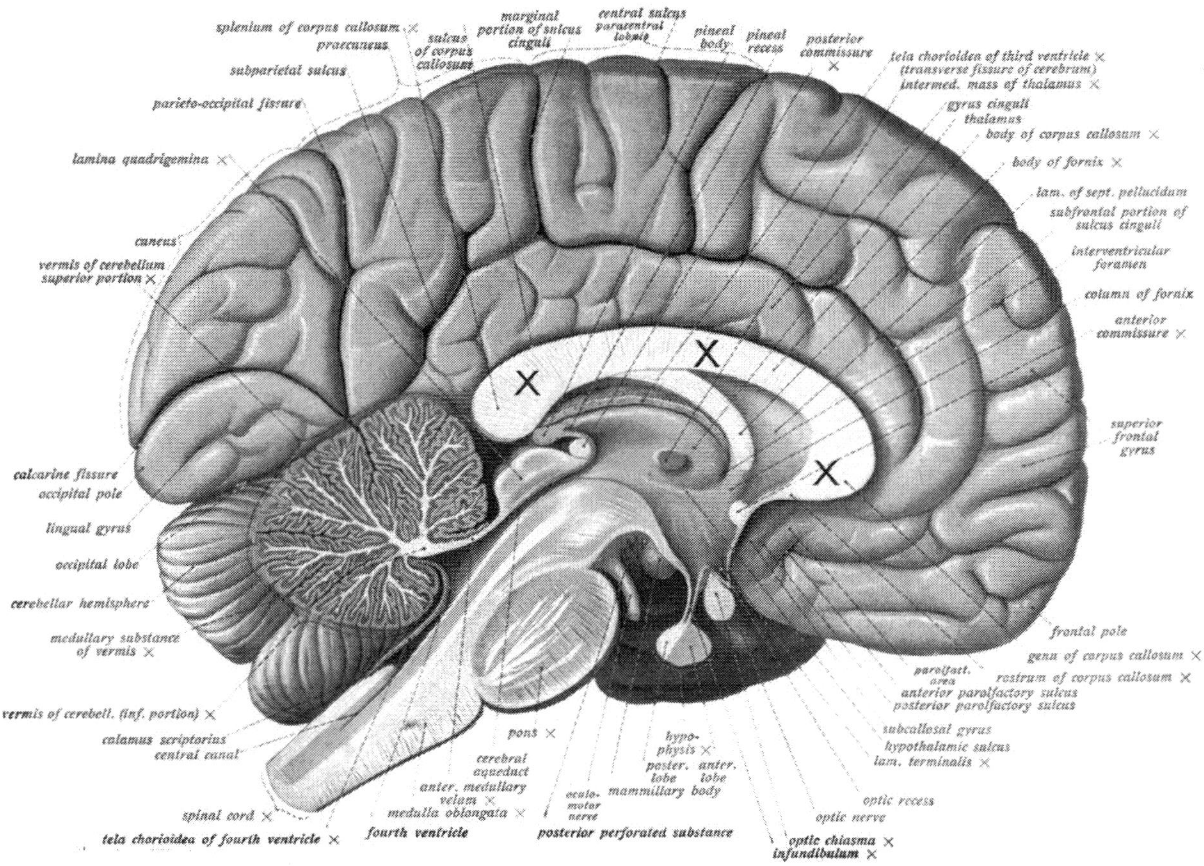

Above is a diagram[1] of a human brain sliced right down the middle – between the eyes and continuing to the back (the sagittal perspective). This diagram is of the left hemisphere – the right hemisphere mirrors its appearance. The Corpus callosum, which connects both hemispheres, is the white element with three large 'X's in the diagram above.

The number of neurons in the human brain was an educated guess until recently. Dr. Suzana Herculano-Houzel and her group have been counting the number of neurons in mammalian brains using an unorthodox scheme. The brain is placed into a cell wall dissolving fluid. Then, the nuclei are stained, and the liquid stirred well. A quantity of liquid is measured out. Counting the number of neurons in the measured quantity of liquid and multiplying by the total volume of liquid gives the number of neurons in

---

[1] Diagram from: Atlas and Text-book of Human Anatomy Volume III Vascular System, Lymphatic system, Nervous system and Sense Organs, by Dr. Johannes Sobotta, 1908, in the Public Domain.

the entire brain.[2] We now know that there are 80 Billion neurons in a typical human brain. Amazingly, at birth, we have all the neurons we will ever have[3].

The next sections are a top-down description of the elements needed to understand the Connections Model.

> <u>Neocortex</u> – A component of the brain and the main subject of the Connections Model
> <u>Neurons</u> – The smallest components of the nervous system, including the Neocortex
> <u>Synapses</u> – The gap between neurons
> <u>Firing</u> – The conditions that lead to a neuron firing into a synapse

But first, an introduction to the general organizational structure of the neocortex.

Hierarchical Organization of Cognitive Function in the Cortex

"The Global Landscape of Cognition: Hierarchical Aggregation as an Organizational Principle of Human Cortical Networks and Functions[4]" is a meta-analysis combining the results of many individual research efforts. The results show a hierarchical progression of increasingly abstract cognitive functions.

From the Executive Summary:

> This study revealed an objective hierarchical landscape of cognition in the brain, with awareness at its structural core, and all results defined solely by computational analysis, largely devoid of human bias. Specifically, our results illustrated two interrelated findings: 1) a structural network hierarchy based on cortical architecture, starting at sensory cortices, and extending to distant regions least connected to sensory input, and 2) a corresponding hierarchy of cognitive functions stretching from primary sensory relay and processing to higher cognition.

Again, from the paper:

> From tangible sensory inputs, symbolic content may progressively emerge as information is processed deeper into the brain's structural network, starting with inputs, and expanding in abstraction or refinement, resulting in intangible or deep symbolic content in the structural pinnacle of the human brain network.

It is important to recognize that the key difference between our central nervous system and other mammals is the dramatic increase in our Association cortex. It is that increase that allows us to have increasingly abstract representations that are the basis for the planning and execution of complex actions that exceed the cognitive capacity of other animals.

---

[2] The Human Brain in Numbers: A Linearly Scaled-up Primate Brain https://www.ncbi.nlm.nih.gov/pmc/articles/PMC2776484/

[3] Aside from a particular region of the hippocampus, the dentate gyrus, which gains neurons as we age, although the growth slows down with age.

[4] The global landscape of cognition: hierarchical aggregation as an organizational principle of human cortical networks and functions | Scientific Reports-2015/12/16/online | Taylor, P., Hobbs, J. N., Burroni, J., Siegelman, H. T. | https://doi.org/10.1038/srep18112.

The Neocortex

The Neocortex, as shown in the diagram below, is the outermost portion of the Cerebral cortex. Other components of the Cerebral cortex are beneath the Neocortex and include the Hippocampus, Amygdala, Claustrum, and Basal Ganglia.

The Neocortex is the largest part of the Cerebral cortex. If you were to unfold it and turn it into a square, it would be about the size of a dinner napkin, 17 to 21 inches[5]. It is thin, varying from .08-.16 inches (2 - 4mm).

While the other components of the brain vary in structure and organization, the neocortex is remarkably uniform. This highly uniform processing engine performs a wide array of diverse functions, include, speaking, reading, singing, dancing, recognizing objects, and people.

The uniformity of the structure of the neocortex suggests that the operating principles of the neocortex are uniform as well. The uniformity should make it easier to understand how it accomplishes these diverse tasks.

Divisions of the Neocortex – by Lobe

As shown below, there are four lobes of the Neocortex, i.e., Frontal, Parietal, Temporal, and Occipital Lobe. Within each lobe are regions that perform specific cognitive functions.

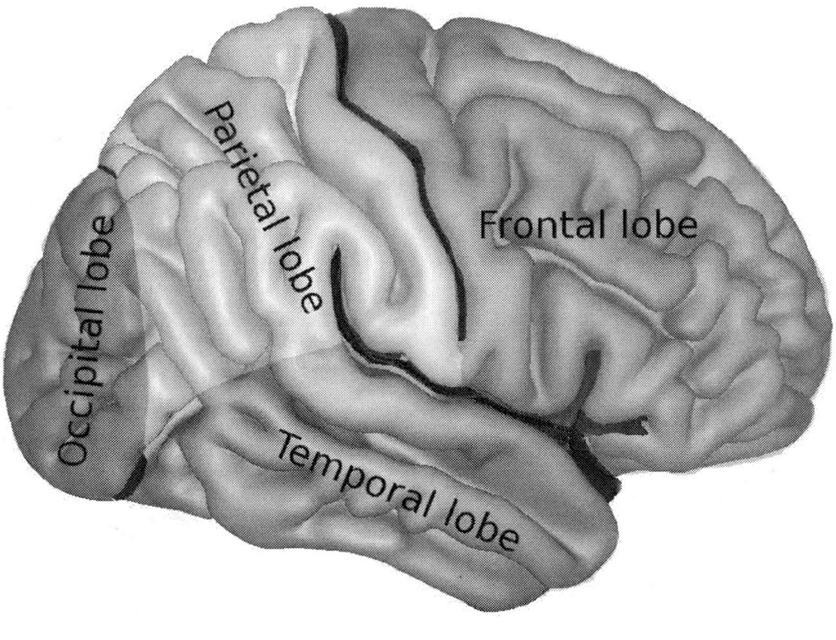

---

[5] Derived from: Brain Size and Folding of the Human Cerebral Cortex, at https://academic.oup.com/cercor/article/18/10/2352/384745

Divisions of the Neocortex – By Function

The figure below illustrates the interrelations between major components of the neocortex.

Receives and categorizes sensory input: e.g., sight, sound, smell, touch, pain, and body position.	Records events and results to guide future behavior. Determines appropriate response	Initiates motor (muscle) activity (including speech)

The Association cortex of humans is the largest of all mammals. It is where we execute our advanced cognitive skills.

Divisions of the Neocortex – By Layer

Neuroscientists have identified six principle layers (vertical) in the neocortex. Each layer has distinctive patterns of connections within and outside of the neocortex. The layers are designated L1, L2, L3, L4a, L4b, L5a, L5b, L6a, and L6b.

An early pioneer of neuroanatomy, Santiago Ramon Cajal, drew detailed representations of stained neural structures, including the figure below[6]:

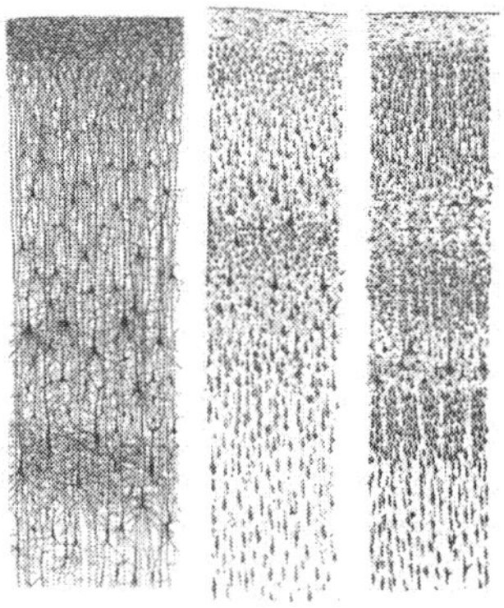

[1] [2] [3]

[1] Visual Cortex of Adult Human
[2] Motor Cortex of Adult Human
[3] Dendrites and Axons from an infant of 1.5 years

---

[6] "Comparative study of the sensory areas of the human cortex" by Santiago Ramon y Cajal, published 1899, ISBN 9781458821898. In the Public Domain. Note the drawings were obtained after Cajal employed Nissl and Golgi stains, which bring out the details shown.

Divisions of the Neocortex – By Area

A recognized organizing structure of the neocortex is the cortical minicolumn. Important details include:

1. Neurons within the minicolumn receive common inputs, have common outputs, are interconnected, and may well constitute a fundamental computational unit of the cerebral cortex.

2. Minicolumns comprise perhaps 80–120 neurons, except in the primate primary visual cortex (V1), where there is typically more than twice the number. There are approximately 2x108 minicolumns in humans.

3. From calculations, the diameter of a minicolumn is about 28–40 μm.

Pyramidal and Cortical Interneurons are the predominant neurons in the Neocortex. Pyramidal neurons stretch vertically through the layers of the neocortex; cortical interneurons tend to stretch horizontally.

Next – Neurons, the building blocks of the Nervous system.

Neurons

Neurons are the fundamental units of the nervous system. They are electrically excitable cells that transmit and process information. There is considerable variation in neuron structure and function.

Dendrites and Axons are common features of neurons. Dendrites receive input signals; Axons send out signals to neighboring neurons. A typical neuron is pictured below[7].

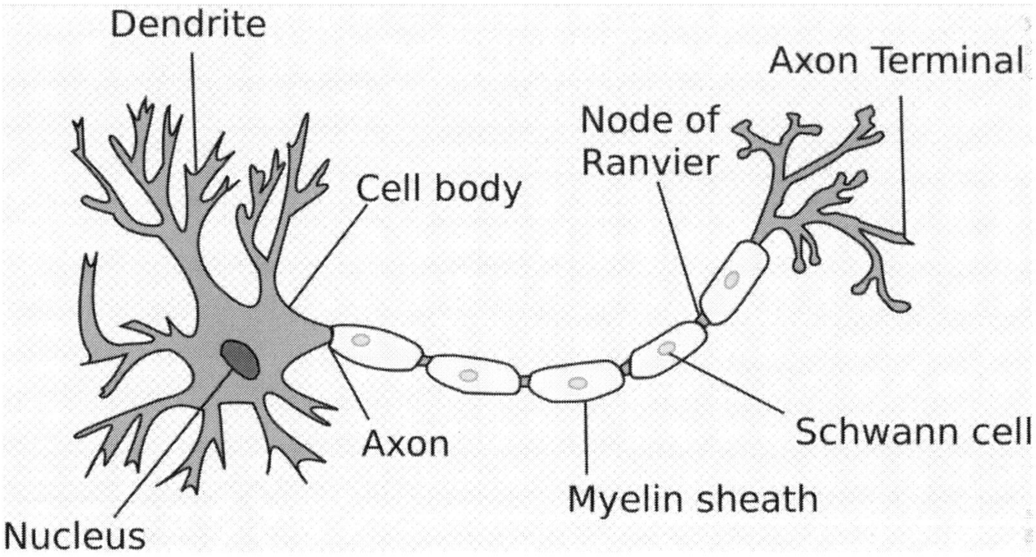

Axons send out neurotransmitters (electro-chemical messengers) that Dendrites receive.

Both Dendrites and Axons may be extensively branched. This branching means that they make contact with many, many other neurons.

Sensory and Motor neurons have a modified structure. In a Sensory neuron, the dendrite is replaced by a component that converts light, sound, pressure into neural signals. In Motor neurons, the axons terminate on muscle fibers or glands.

As mentioned previously, in the Neocortex, there are two principal types of neurons: Pyramidal and Cortical interneurons. Descriptions follow.

---

[7] Above graphic by Quasar Jarosz at English Wikipedia, CC BY-SA 3.0, https://commons.wikimedia.org/w/index.php?curid=7616130

Pyramidal Neuron

Pyramidal neurons are workhorses of the Neocortex. A Pyramidal neuron and its major components[8] are pictured below.

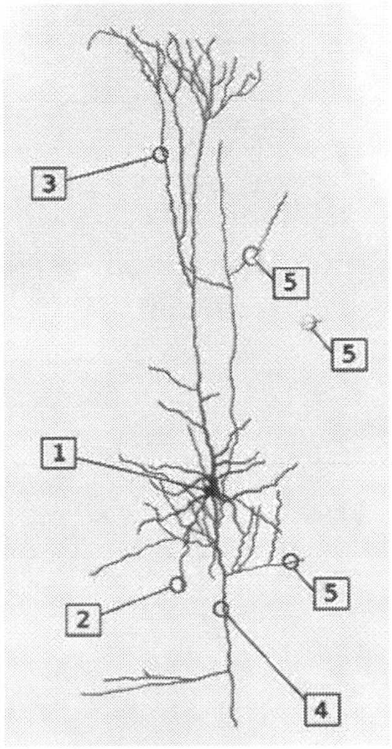

1) Soma – shaped like a pyramid – the cell body

2) Basal dendrite - arise from the base of the soma, branch profusely

3) Apical dendrite - arise from the apex/top of the soma, extends towards the cortical surface

4) Axon – descends – often reaches outside the minicolumn

5) Collateral axon – branch off the main axon

Dendrites of a pyramidal neuron are in the same minicolumn as the pyramidal neuron. Their Axons, on the other hand, escape the minicolumn and can be up to 1 meter in length. These axons terminate in other minicolumns within the neocortex or in regions outside of the neocortex. The axons of neurons from outside the neocortex can reach into the neocortex as well.

A single pyramidal cell has nearly 30,000 Dendritic spines, which receive excitatory inputs, and approximately 1,700 spines which accept inhibitory inputs.

The number of minicolumn destinations that an axon can reach is harder to determine, as the axons reach far beyond their immediate vicinity. There are estimates that there can be up to 10,000 branches in the axon of a Pyramidal neuron. In *Patterns of Axon Collateralization*[9], the authors identify axons that split into 2-5 axon branches, running for several millimeters, branching again to enter into multiple neural minicolumns. In *"The Quantitative analysis of axon collaterals*[10]*,"* the authors estimate that individual pyramidal cells in just one of the layers of the neocortex, layer IIb, form contacts with up to 3,000 other pyramidal neurons.

In the Connections Model, Axon branching is key. Extensive branching gives the neuron flexibility in where it will establish a synapse in the future.

---

[8] Sourced from Wikimedia commons, author listed as Fabuio.

[9] Ojima, Hisayuki & Honda, Christopher & G Jones, E. (1991). Patterns of Axon Collateralization of Identified Supragranular Pyramidal Neurons in the Cat Auditory Cortex. Cerebral cortex (New York, N.Y.: 1991). 1. 80-94. 10.1093/cercor/1.1.80.

[10] Quantitative analysis of axon collaterals of single superficial pyramidal cells in layer IIb of the piriform cortex of the guinea pig. |ul Quraish A1, Yang J, Murakami K, Oda S, Takayanagi M, Kimura A, Kakuta S, Kishi K.

Cortical Interneurons

The other principal neuron in the Neocortex is the Cortical interneuron. Unlike Pyramidal neurons, Cortical interneurons generally traverse small distances within a single cortical minicolumn. They may terminate on dendrites, the beginning portion of the axon, or on the cell body of a neuron (called the soma). If they have an axon, it is short. Some terminate on other interneurons.

Most Cortical interneurons are inhibitory – meaning that their actions tend to inhibit the firing of pyramidal neurons. Interneurons apply the brakes, preventing a storm of firings from overwhelming the neocortex.

A Chandelier Interneuron[11] appears below. Chandelier interneurons can connect with multiple different neurons. Note that the cell body or Soma is at the center, the lighter elements on top are the dendrites, and the darker elements below are the synapses.

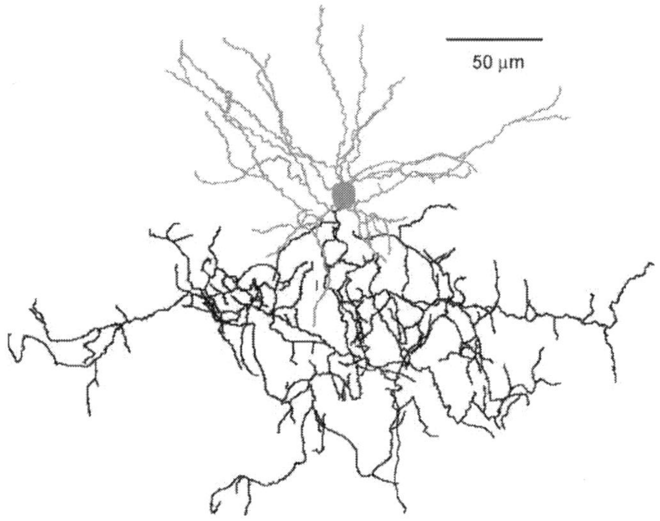

The space between the axon of one neuron and the dendrite is the synapse. Their description follows.

[11] Sourced from, used in the Wikipedia article on Chandelier cortical interneurons. Originally from Woodruff A, Yuste R (September 2008). "Of Mice and Men, and Chandeliers". PLoS Biol. 6 (9): e243. doi:10.1371/journal.pbio.0060243. PMC 2553849. PMID 18816168

The Synapse

Neurotransmitter signals in the nervous system travel from the axon of the sending neuron into the dendrite of the receiving neuron. The small gap between the sending neuron and the receiving neuron is the Synapse.

The sending neuron is also called the presynaptic neuron. The receiving neuron is also called the postsynaptic neuron. In the diagram below, the presynaptic neuron is on top, and the postsynaptic neuron is below. The synapse is in between the presynaptic and postsynaptic neurons.

Neurotransmitters from the Presynaptic neuron traverse the synapse and bind to the postsynaptic neuron. The neurotransmitters raise the voltage of the postsynaptic neuron. Then, they travel back and to the presynaptic neuron, where they are absorbed.

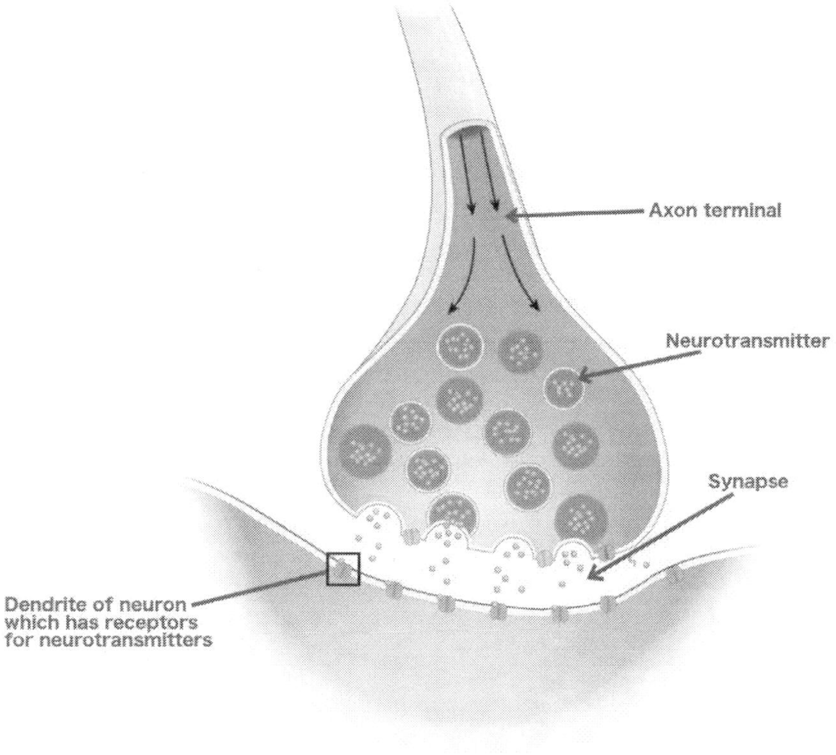

In early brain development, exuberant synaptogenesis is responsible for creating a large number of synapses. Afterward, synapse creation slows dramatically.

Creating Synapses (after Exuberant Synaptogenesis)

Pyramidal neuron dendrites have small protrusions, called dendritic spines. New synapses form between dendritic spines and nearby axons. Dendritic spines continuously change their shape. When adjacent axon's fire, the shape of the dendritic spine changes – moving it closer to the firing neuron – assisting in the creation of a new synapse. Pictured below are example shapes (splines) of dendritic spines[12].

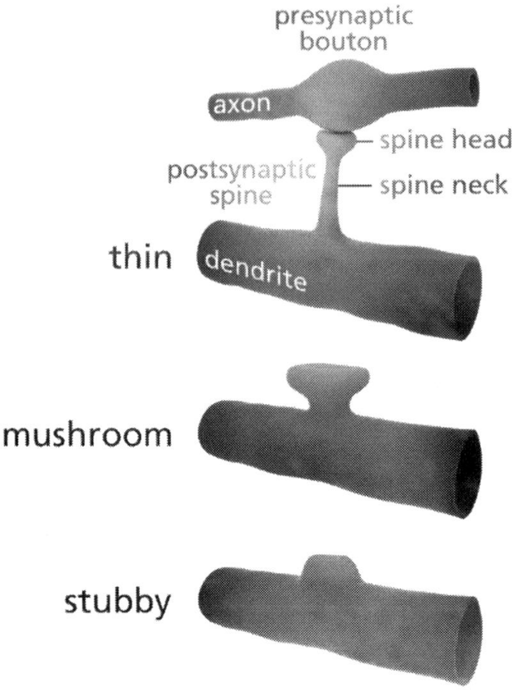

Strength of a Synapse

The strength of a synapse, or Synaptic Strength, refers to the efficiency of a synapse. It is measured by the amount of current in the postsynaptic neuron after the presynaptic neuron fires. The stronger the current, the stronger the synapse. Synaptic Plasticity refers to the idea that the strength of a synapse can change.

Repeating something over and over helps you memorize/learn it. Each repetition results in a neuron firing. Each firing increases the strength of the synapse encoding the new information. Stronger synapses last longer, as it takes longer for a strong synapse to disappear.

Conversely, if a synapse does not fire, and is dormant, its strength will decline. When it declines enough, the synapse disappears. The information encoded in that synapse is lost. The common expression for this is "use it or lose it."

In essence, the mechanism for strengthening a synapse is the same as the mechanism that creates the synapse in the first place.

---

[12] Author Thomas Splettstoesser (www.scistyle.com). Licensed under the Creative Commons Attribution-Share Alike 3.0 Unported license.

Connections – Feedforward and Feedback

There are two general directions for signaling in the neocortex:

> Feedforward – connections that run from sensory input, in general towards the front of the neocortex. [Indicated by solid lines below.] AND
>
> Feedback connections run in the opposite direction. [Indicated by dashed lines below]

In the example below, the feedforward connections are how we know that {tomato, celery, kale} are members of the vegetable category. The dashed lines represent the knowledge that the three items are examples of vegetables.

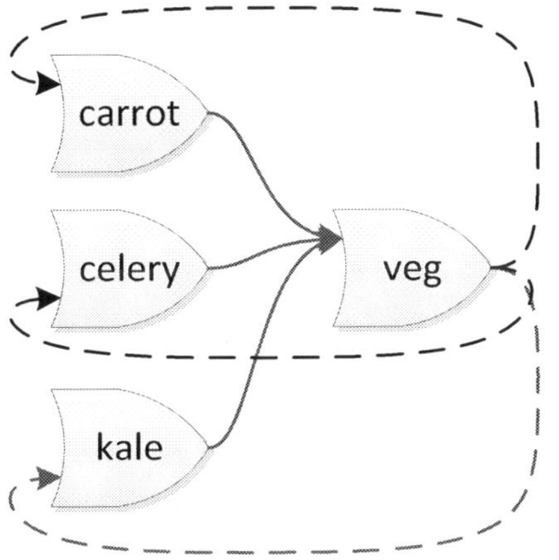

You can reasonably ask, what is the purpose of the feedback connections? Can't you go from vegetable to celery by going backward? Neural connections are one way. When celery fires, it leads to vegetable firing. Vegetable firing doesn't lead to celery firing unless there is a feedback connection.

To illustrate, consider the following questions.

You are taking a test. Is kale a vegetable? The solid connection from kale to vegetable means you can answer the question immediately.

Or, the question is – name a vegetable. With the dashed connections, you can immediately go from vegetable to any of three examples. Without those connections, you would have to search your memory. That might take a while.

Next, neurons fire when their voltage level exceeds a threshold value.

Neurons Fire

The voltage level of the postsynaptic neuron changes as it receives neurotransmitter. Some neurotransmitters increase the voltage level of the receiving neuron, while others decrease the voltage level of the receiving neuron.

Decay

When the voltage level of the neuron exceeds a threshold, it fires. A neuron whose voltage level has been elevated but does not reach its threshold gradually decays back to its resting voltage, as illustrated below.

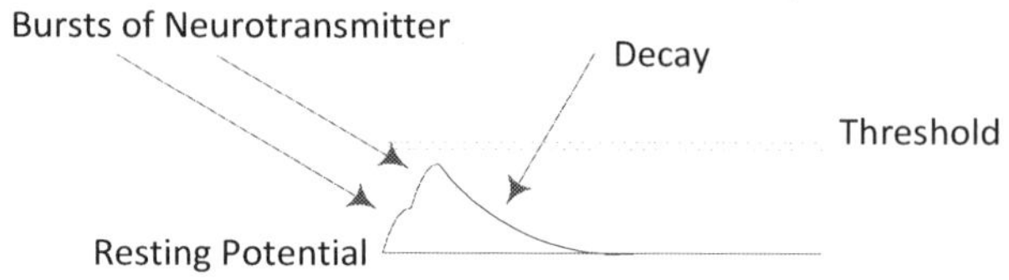

Multiple bursts add together

But don't reach Threshold

Then voltage decays

There are two mechanisms for increasing the voltage level of a neuron.

Temporal summation

Each packet of neurotransmitter changes the voltage level of the receiving neuron by a small amount. Pictured below is the effect of multiple neurotransmitters increasing the voltage level of a neuron until it exceeds its threshold.

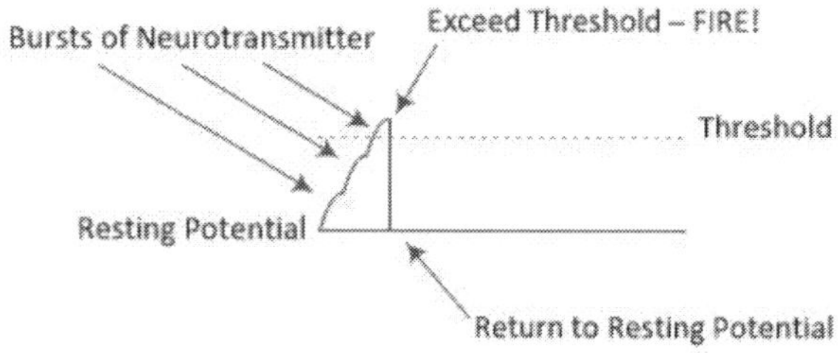

Temporal summation

Many bursts add together

And together exceed the Threshold

Spatial summation

When multiple points surrounding a neuron receive neurotransmitters simultaneously, the change in voltage of the neuron is the sum of the individual voltage changes. When the voltage level exceeds the threshold, the neuron fires.

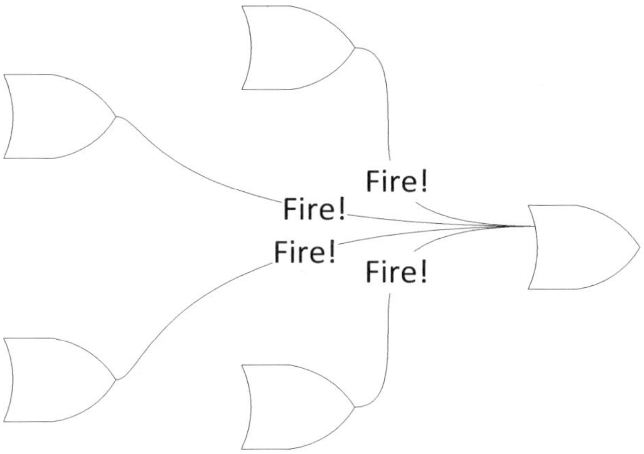

Both of these mechanisms can help create a new synaptic connection or strengthen an existing one.

Inhibitory Neurons

Inhibitor neurons lower the voltage of the receiving neuron. In the Neocortex, Cortical interneurons are often inhibitory. Without mechanisms to lower a neuron's voltage, a cascade of firing would overwhelm the brain.

Concurrent

All of these – temporal summation, spatial summation, inhibition occur simultaneously. When the combined impact exceeds the threshold value, the neuron fires.

The Connections Model

Diagram Symbols

This model that follows describes the conditions under which synapses form and strengthen in the Neocortex. While the mechanisms are simple, the resulting connections enable our cognitive abilities.

Diagrams of neural connections accompany the text. The diagrams are purposefully simple, highlighting a new concept. A full implementation would be much more complex, obscuring the newly introduced concept.

These are the symbols used in the neural connection diagrams.

Pyramidal Neurons

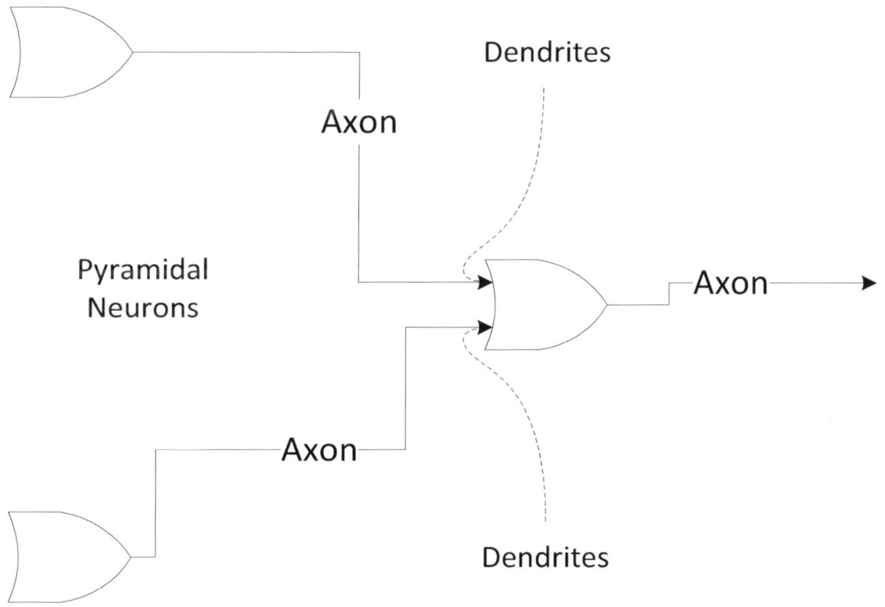

Cortical Interneurons

The cortical interneuron is the object with the word NOT inside.

Strengthening Synapses

When a neuron's axon fires, no connection is created between that neuron and the neurons downstream of it.

But, when two neurons fire at the same time, or nearly the same time, the strength of the synaptic connection will increase.

> <> When they fire at the same time, the effects add - Spatial summation. With each paired firing, the synapse strengthens. When the synapse is strong enough, the downstream neuron will also fire when both upstream neurons fire.
>
> <> When they fire at nearly the same time, their effects add as well - Temporal summation. With each paired firing, the synapse strengthens. When the synapse is strong enough, the downstream neuron will fire when the first neuron fires. However, the second neuron must fire, or else the strength of the synapse will decline.

One other well-known mechanism for strengthening synaptic strength is "Spike Timing Dependent Plasticity" (STDP). STDP requires that the two neurons fire within 50 milliseconds of each other. In the Connections Model, STDP is used to determining the speed of movement across a visual field. Elsewhere, the tight 50-millisecond requirement restricts general applicability and is the main point of "Beyond STDP – towards diverse and functionally relevant plasticity rules[13]."

> One key difficulty with invoking STDP-like plasticity rules to support all forms of associative learning is the difference in timescale between plasticity rules, at the scale of tens of milliseconds, and the contingencies of behavioral learning, often on the timescale of minutes.

The main mechanisms of the Connections Model involve two events that do not require the tight time constraint of STDP. Both involve two events.

> <u>Coincident</u>: When two events always occur together at the same time, the firing of a single neuron represents their correlation (based on spatial summation). They are both consequences of a common, underlying event – identified as 'x' below.
>
> <u>Prediction</u>: When one event always precedes the other event, we conclude that the first event caused, or at least predicts the second event (based on temporal summation).

In the Coincident circumstance, not only must there be a neuron that receives the signals, the signal from both must arrive at the same time.[14]

---

[13] Beyond STDP - towards diverse and functionally relevant plasticity rules; Aparna Suvrathan; https://doi.org/10.1016/j.conb.2018.06.011

[14] There are mechanisms that adjust the speed of transmission down axons that facilitate coincident firing, and speed up future firing.

In summary:

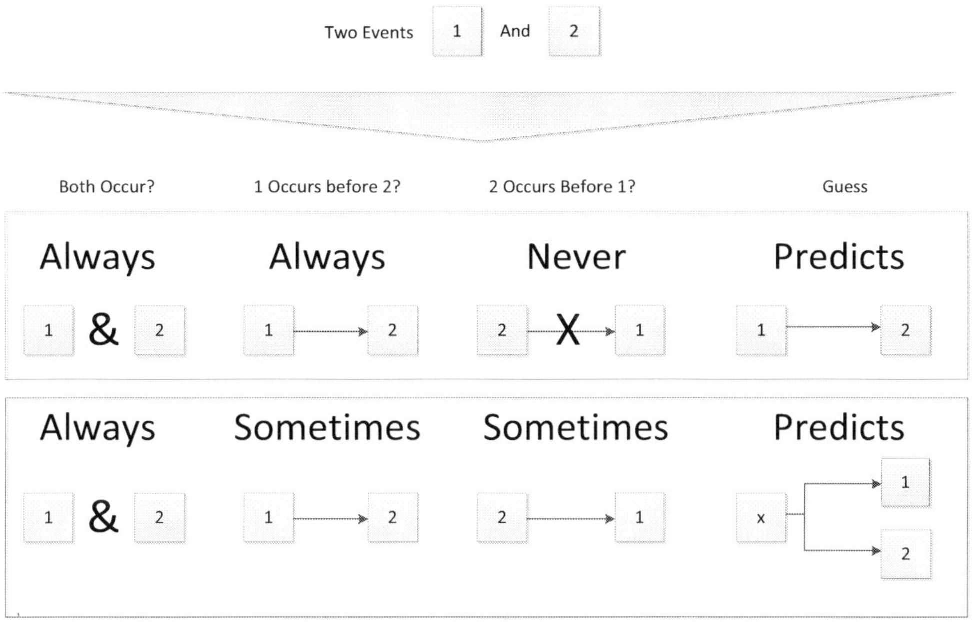

A sneak peek at what is to come

The perceptual process is the mechanism by which we recognize commonly encountered objects. Its foundation is the recognition that two individual elements (or events) are part of the same element. A hierarchy, of such recognitions, results in the identification of a commonly encountered object.

Prediction is the process of recognizing that one thing always precedes another. Temporal summation drives prediction. The firing of the first neuron warns us of what is to come, giving us a chance to avoid danger.

Both of these processes are fundamental to the Connections Model.

Classical Conditioning

The prediction mechanism for creating new synapses underlies Classical Conditioning. A popular description of Classical Conditioning involves a bell, dog food, and a hungry dog. The left flowchart below is the process to condition the dog's reaction to the bell.

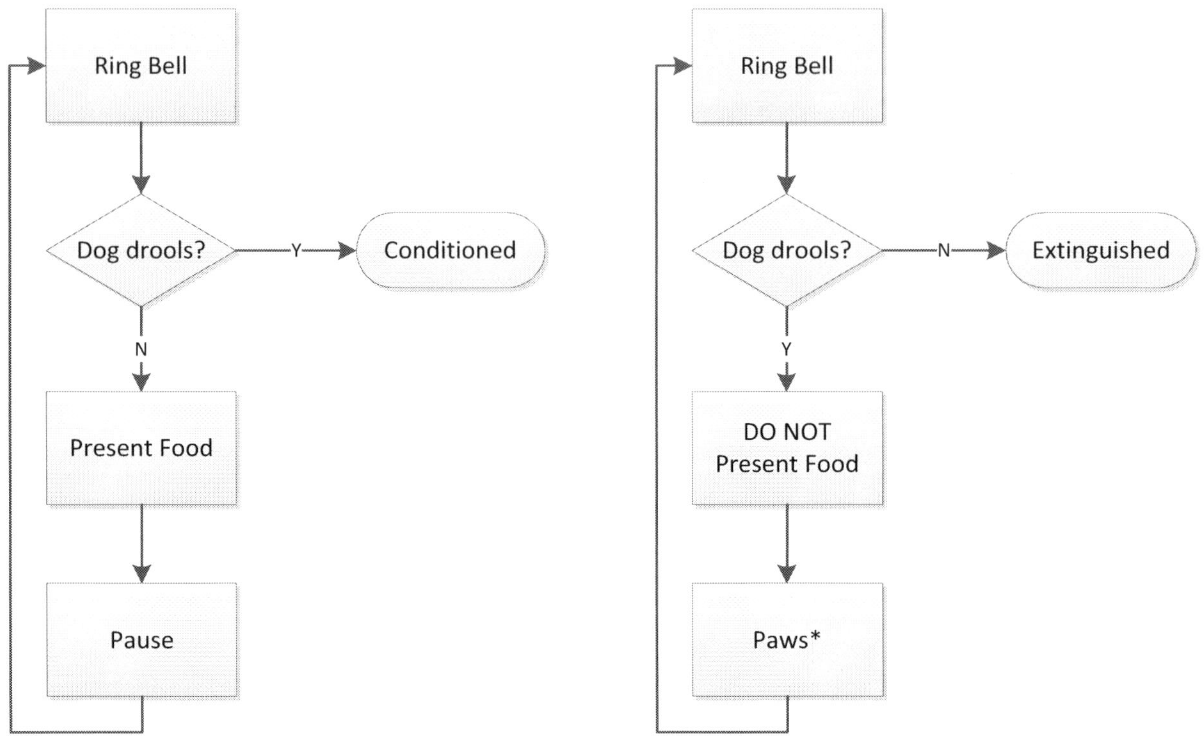

Conditioning and Extinguishing of Classical Conditioning

The flowchart on the right above is the process that extinguishes or eliminate the dog's reaction to the bell.

But, after extinguishing, reconditioning is quicker. Fewer trips around the loop are required to recondition the dog to drool in response to a bell.

Hence, even one conditioning brings about a permanent change to the conditioned synapse. The permanent change explains why it is easier to relearn a forgotten detail than to learn it the first time.

*You might note the replacement of the word 'Pause' with the word 'Paws' in the flowchart on the right. Yes, it is an attempt at humor.

Classical Conditioning Neurological Connections

Now, let us trace the formation of the connection that results in the dog drooling when presented with food. There are three neurons involved.

Food – Fires when Food is in front of the dog.
Bell – Fires when the bell rings.
Drool – When it fires, the dog drools.

Before conditioning, there is a connection from the presentation of food to the motor neurons that lead to drooling, i.e.

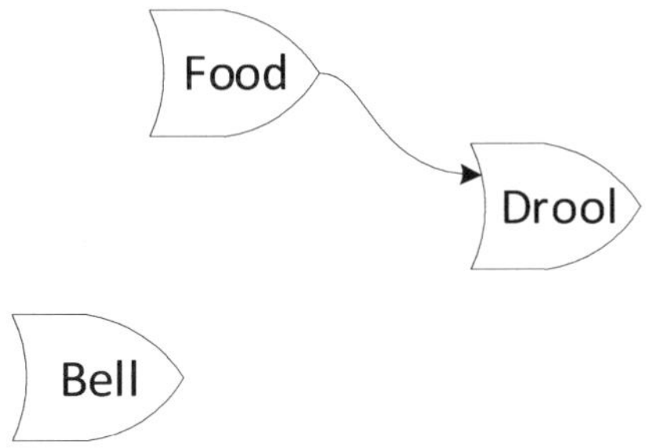

One round of conditioning looks like this:

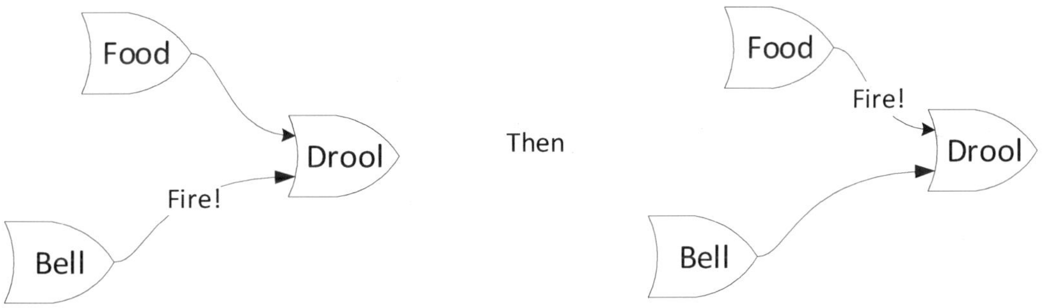

The prediction mechanism, based on temporal summation, leads to the creation of a synapse from the Bell neuron to the Drool neuron. Repeated ringing the bell followed by food strengthens the new synapse.

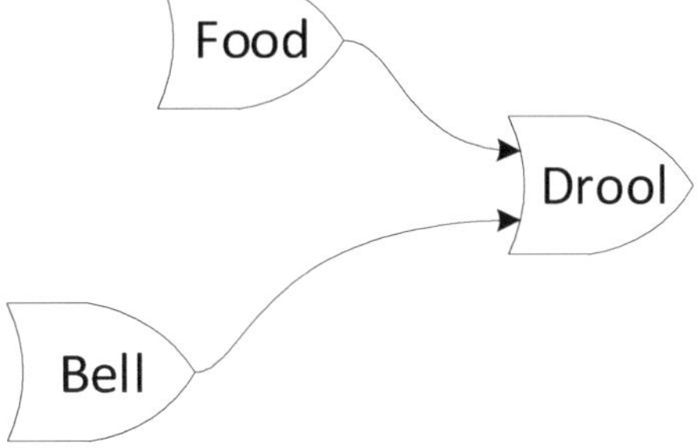

There are two issues. The time between the bell firing and the presentation of food is too long. When the Bell neuron fires, it raises the voltage of the Drool neuron. But the Drool neuron will decay to its resting state well before the Food neuron fires.

Somehow, the Drool neuron has to retain its elevated state long enough for the Food neuron to fire and push it over its voltage threshold. The mechanism for this is described in the section below, 'Hold that thought.'

Another problem is – how is it that an axon endpoint of the Bell neuron terminates on the dendrite of the Drool neuron? That seems unlikely. Recall that the axons of a pyramidal neuron branch out profusely. For the bell ringing to lead to a dog drooling, there must be a connection between the two neurons. Perhaps the reason for the profuse branching of pyramidal neurons is to allow this kind of connection? The branching is required as the neuron can't know which connection will be useful in the future.

Survival Advantage

The process above where one event predicts another helps improve your chance of survival.

Imagine you hear a rustling in the grass. You know that the rustling in the grass could indicate that a tiger is nearby. You take no chances and run away. Another person also hears the rustling in the grass but does not realize what it can mean. Which one of you does the tiger eat?

You might have learned that rustling in the grass may indicate a tiger is nearby, or you might have encountered rustling in the grass and escaped a tiger. Either way, you run away!

Is this an overreaction?

Perhaps the probability is only 10%, meaning, 90% of the time, there is no tiger. But after one experience where a tiger was present, you take no chances. Because of this, you will survive a tiger that is nearby.

Individuals who do not run from rustling in the grass will be eaten by the tiger 1 out of 10 times.

This phenomenon, of being over-cautious, is part of human behavior. Common sense tells us that it is "better to be safe than sorry." The truth is, we are the survivors who over-reacted to a potential threat.

Correlation – Simultaneity of Events

In the diagram below, neuron A and neuron B consistently fire at the same time. There is at least one location where a neuron, labeled C below, receives signals from the axons of neurons A and B[15].

Neuron C is a coincidence detector. When neuron A and neuron B both fire at the same time on neuron C, spatial summation occurs. The result is two new synapses; from neuron A to neuron C and from neuron B to neuron C. Repeated simultaneous firing of neurons A and B strengthen the new synapses.

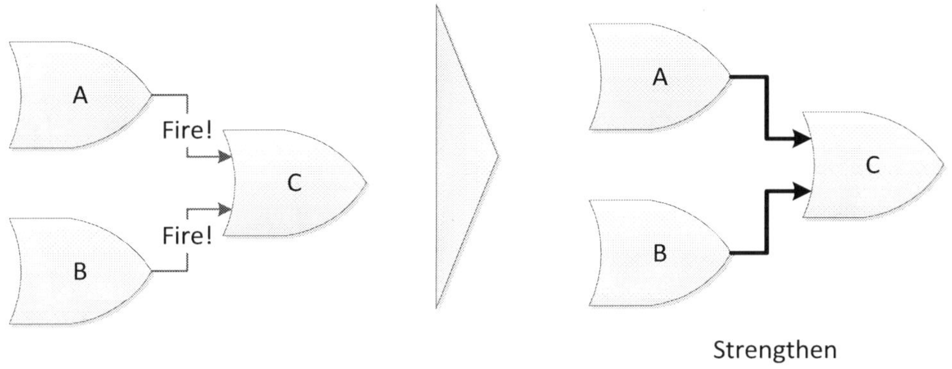

Repeated coincidental firings of neurons A and B strengthen the synapse. If the initial coincidence did not repeat, the synapses would fade away. Spatial summation is the synaptic mechanism involved.

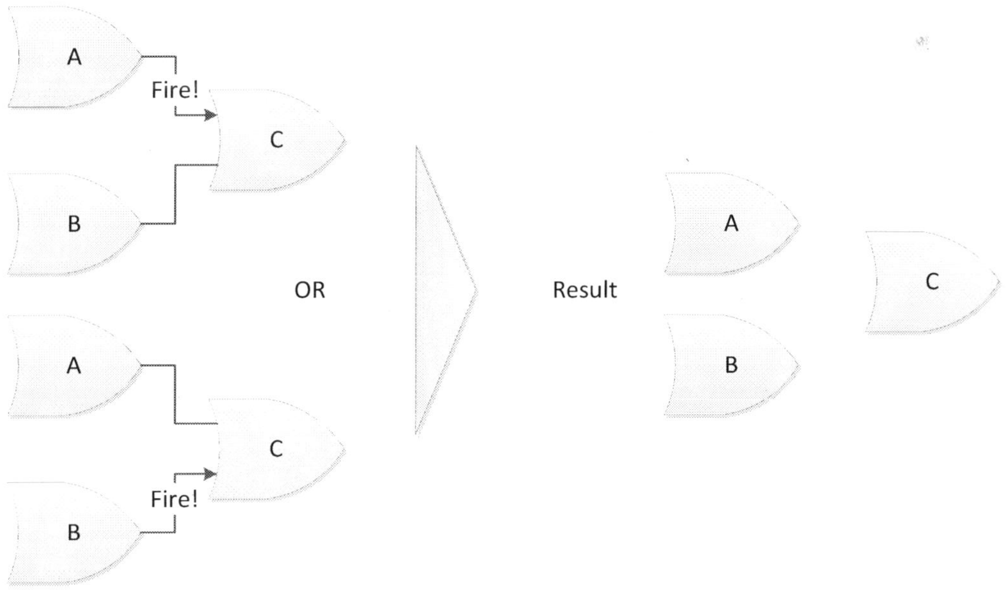

15 There may be more than one neuron that both neuron A and neuron B synapse upon. This adds redundancy.

Likelihood of available neuron for Correlation

How likely is that neuron A and neuron B both send axons to a common minicolumn with neuron C?

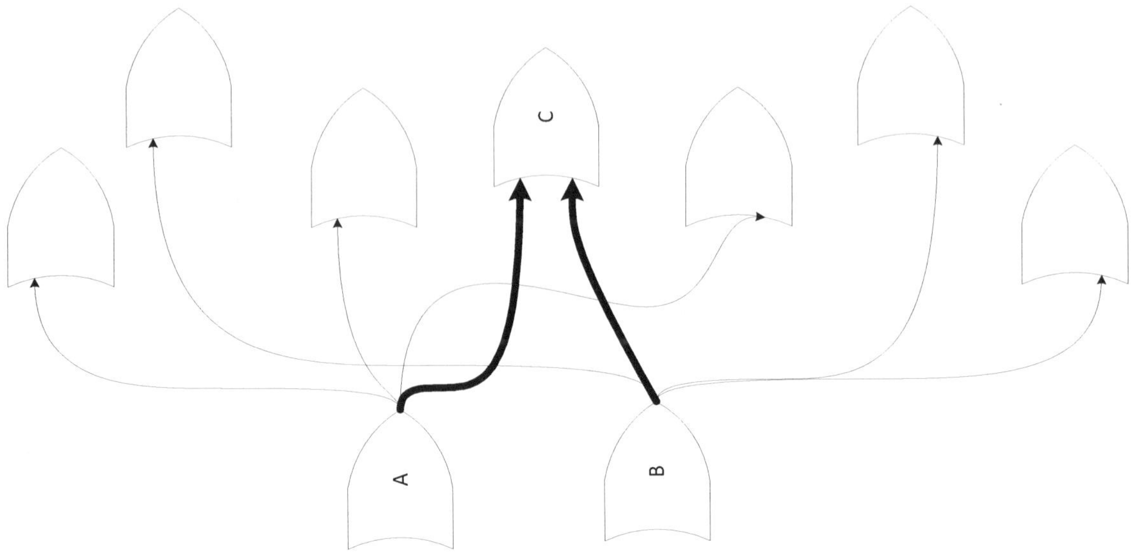

Axons of a single pyramidal neuron in the neocortex branch profusely, terminating in many different minicolumns. Perhaps the reason for this branching is to make possible this seemingly unlikely coincidence.

Neurons A and B are comparatively close to each other, residing in the same functional region. Both send out axons that terminate in 10,000 different minicolumns. The likelihood that there is a common neuron, neuron C, in the same functional region, reached by both is high.

See these two papers for support of this hypothesis.

> "A compact and realistic cerebral cortical layout derived from pre-whitened resting-state fMRI time series: Cherniak's adjacency rule, size law, and meta-module grouping upheld." Frontiers Neuroanatomy, 06 September 2012 | https://doi.org/10.3389/fnana.2012.00036.
>
> "Global optimization of cerebral cortex layout" Christopher Cherniak, Zekeria Mokhtarzada, Raul Rodriguez-Esteban, and Kelly Changizi; PNAS 2004 January, 101 (4) 1081-1086 | https://doi.org/10.1073/pnas.0305212101

Next Up – Perception

The simultaneity of two neurons firing together always indicates a common cause. Perception builds upon that recognition.

The first topic is one of Prediction-Error, a relatively new idea that says we perceive what we expect, or predict, what we will perceive.

Perception - Prediction Error

A relatively new idea in perception is that you see/hear/smell/taste what you predict or expect to experience.

A key metric of the prediction hypothesis is Prediction-Error. The difference between what you expect/predict you will see/hear/smell/taste and what you observe is Prediction-Error.

One hypothesis is that there is a shortcut from early sensory processing into the Prefrontal Cortex. The Prefrontal Cortex makes a prediction based on this preliminary information. Feedback circuits that return to sensory processing steer your perceptual process.

Real-time information can guide prediction. For example, if you hear a lion's roar behind you, you expect to see a lion when you turn around. Or, when you hear your friend's voice, you expect to see your friend's face. When it is another person, it takes a moment to reset your expectations and see who it is. As the lion roar's example indicates, it can be a survival advantage to anticipate what you are about to experience.

In the visual system, there are multiple, different regions of the neocortex that perform unique visual recognition. One region specializes in recognizing faces, another that specializes in reading letters, and another recognizes body parts. Now, imagine if all of these regions are simultaneously analyzing the visual stream. How would the neocortex determine which one was the one to use?

Instead, a 'prediction' or guess is made and used to direct the visual stream to the most appropriate visual processing system. If the result of the analysis is good, as in, the prediction error is small; the neocortex goes with that result. Otherwise, another prediction occurs.

You can imagine this process as a rotary selection switch.

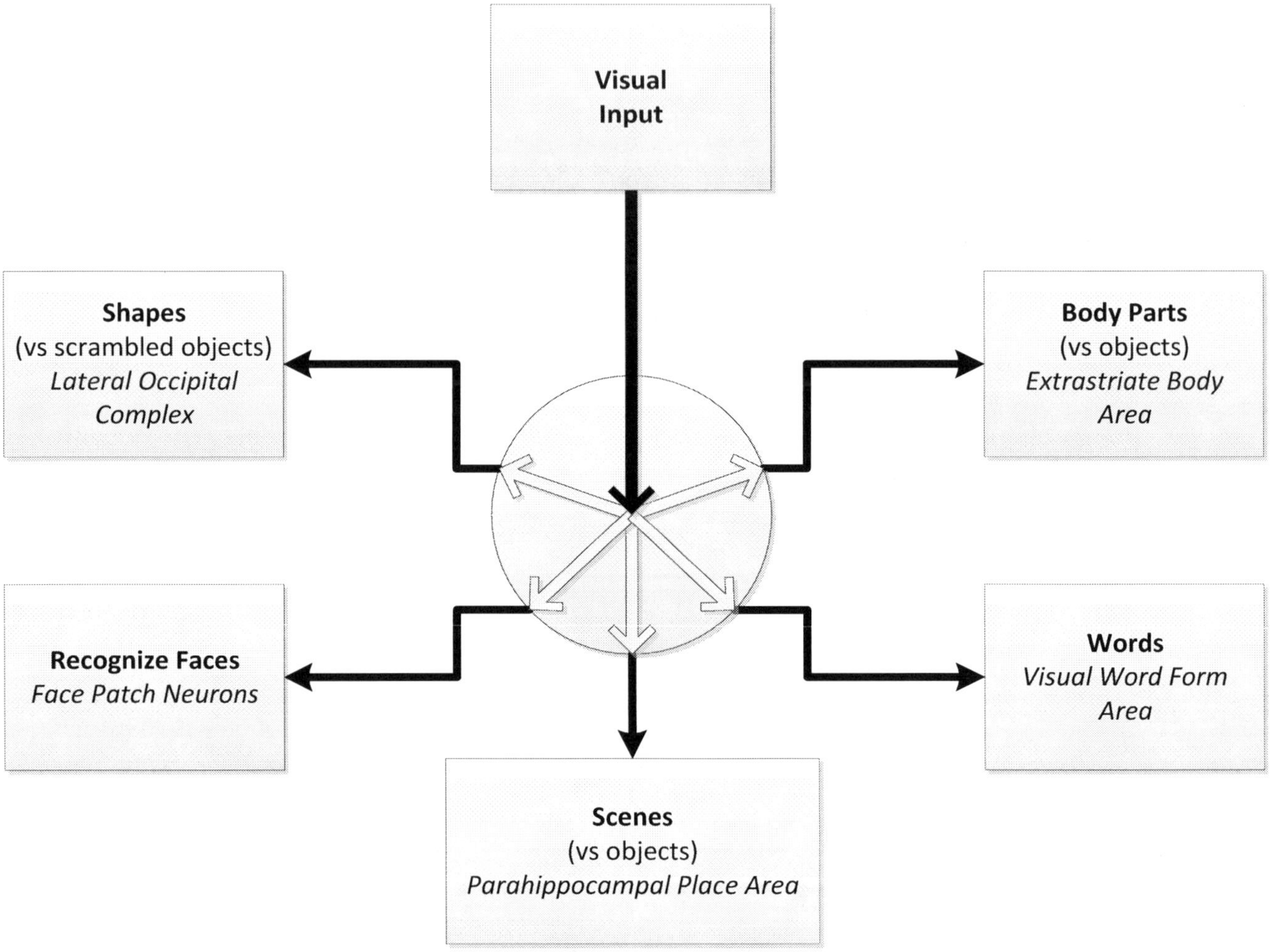

In this diagram, the prediction mechanism turns the switch to the most likely specialty circuit. Gating neurons perform the selection[16].

---

[16] I am not certain that the gating or selection process occurs at the front end, where visual input is directed, or whether the selection somehow occurs at the back end, after the individual recognition process occurs. The above is the more efficient process, making it the more likely one.

Gating Neurons

The term gating neuron comes from the similarity to a gate. Like a gate, when it is open (active), it allows passage. When closed, it prevents passage.

In the gating circuitry below, the **Select** neuron needs to fire continuously to pass the **Input** neurons signal to the **Output** neuron, as per below[17].

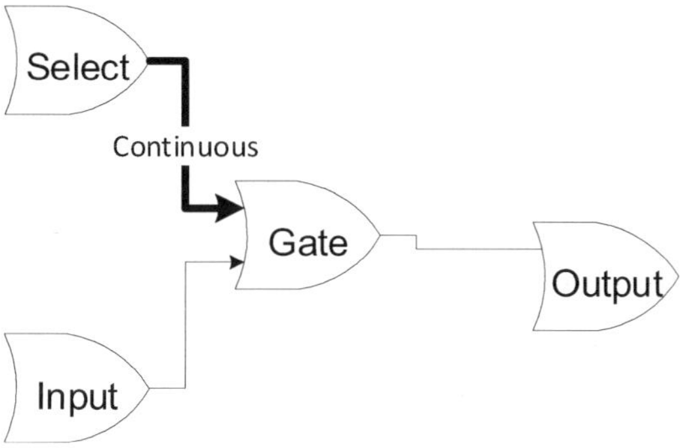

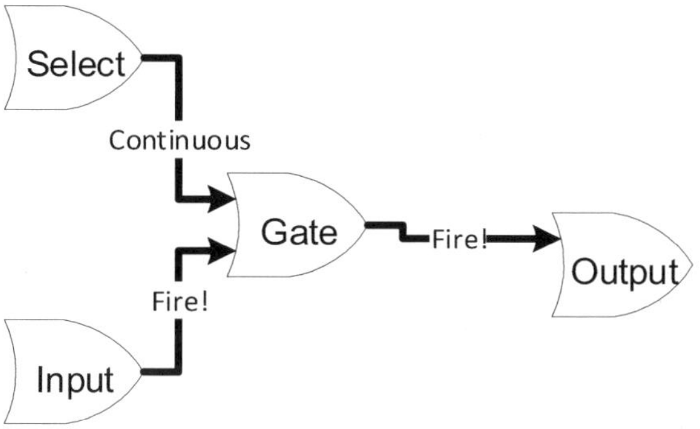

When the **Select** neuron is not active, the **Output** neuron is continuously silent, no matter the signal from the **Input** neuron, as per below. Without a select input, the output will always be silent.

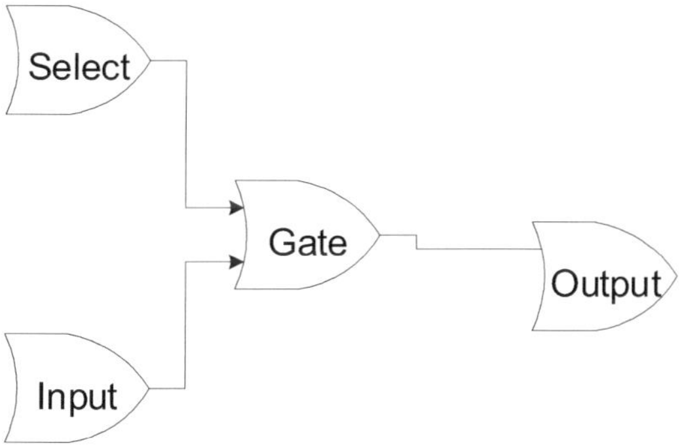

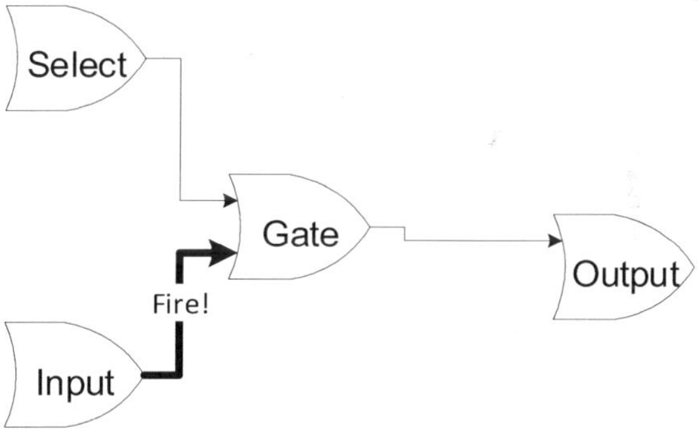

Alternatively, unwanted signals might be suppressed by a cortical interneuron, as per below.

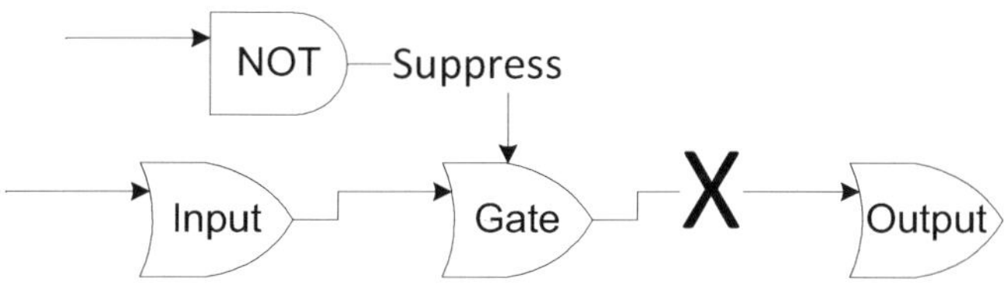

[17] The select neuron might be a Tonically active neuron, which fires more or less continuously, unless a cortical interneuron is inhibiting it. Another possibility is an interneuron suppressing the firing of the Tonically active neuron.

As described above, visual input is directed based on prediction to the expected region for analysis. A set of gating neurons direct the visual field output into each of the specialized analysis systems, corresponding to the rotary switch analog, as shown below.

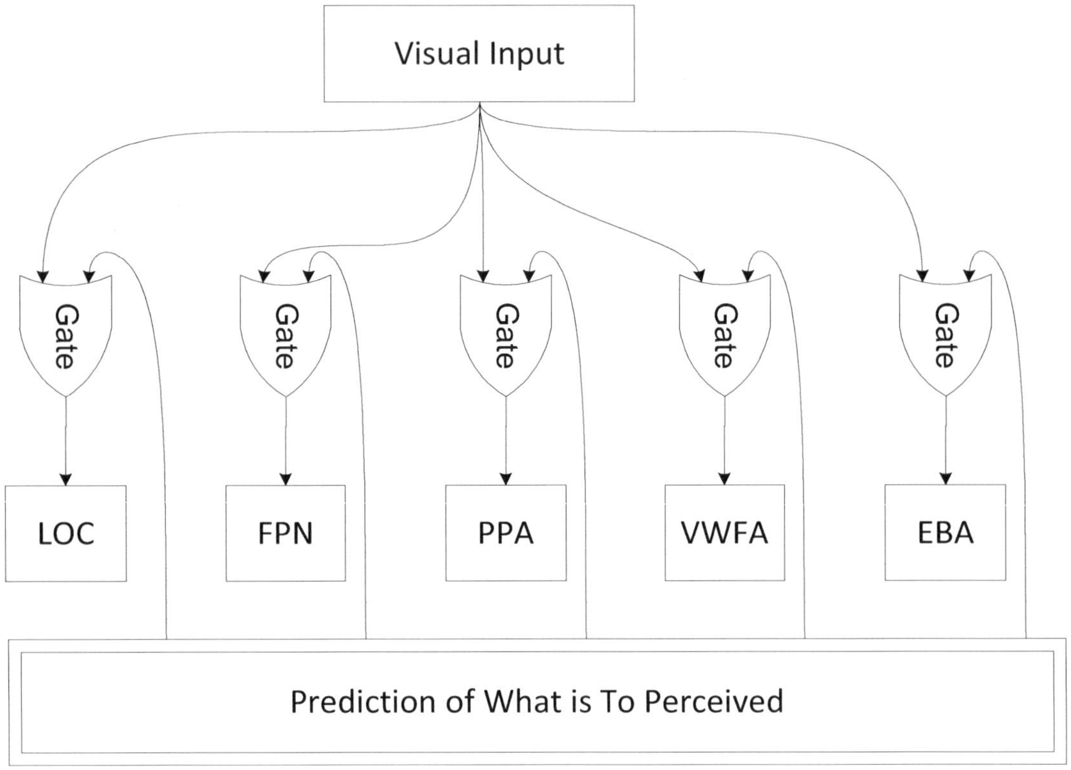

The prediction mechanism does not end here. Prediction guides recognition in subsequent portions of visual (sensory) processing.

Experiments have revealed that a subject will recognize a red dot faster when they are told, in advance, to expect a red dot. Quicker recognition suggests there is a prediction mechanism in place at the level of color detection.

Up until recently, it was a mystery as to why the number of feedback circuits in the visual cortex outnumbered feedforward connections. The gating mechanism is a likely culprit.

Comedy as Prediction Error

When listening to a stand-up comedian, the comedian leads you in one direction, then finishes up with a conclusion from a different direction. Your prediction about what was to come was wrong. You are surprised. The pleasant surprise reaction results in dopamine secretion. Dopamine enhances new synapse formation. The difference between what you expected and what was delivered is prediction error. Pleasant prediction error undergirds the mechanism of comedy.

Perception - Visual Cortex

In the next section on perception, we move beyond the Prediction mechanism into the more customary view of perception, in which each layer of the process recognizes increasingly abstract qualities and concepts.

A high-level view of the Visual cortex is below. Current models of the visual cortex indicate that there is one pathway for recognition of objects, the ventral stream, and another pathway for the recognition of motion and location, the dorsal stream.

Intriguingly, a component of the Cerebral cortex, the Hippocampus, is involved in keeping track of where you are. The dorsal stream likely sends output to the Hippocampus.

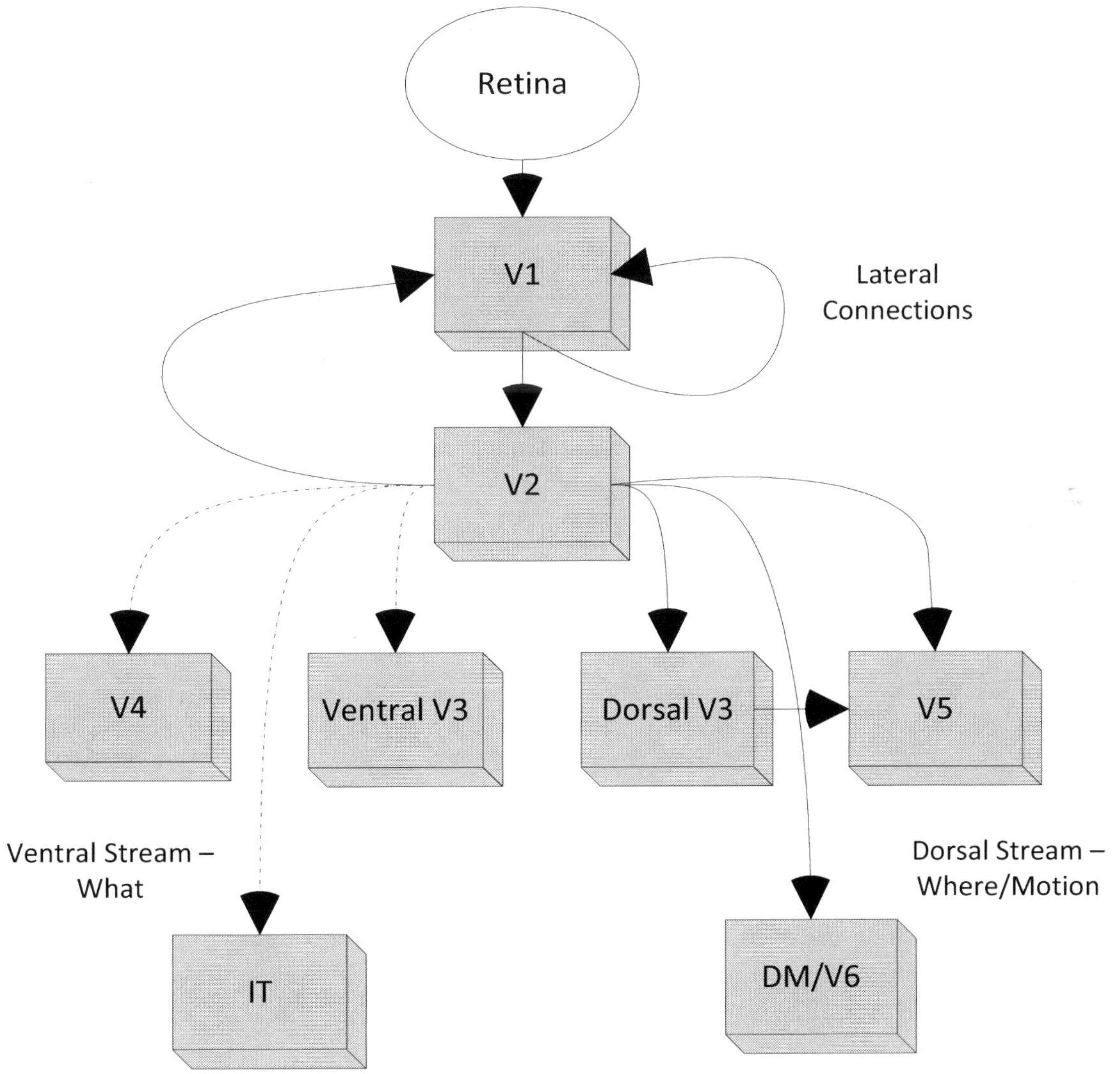

Perception - Recognizing Objects (Ventral Stream)

Perception recognizes simultaneously occurring events/neural firings

Perception is a process that makes sense out of the incoming flood of sensory input. Visual perception recognizes repeating patterns in the input stream. The process builds, as subsequent layers identify:

1. Lines and edges
2. Shapes
3. Objects

The mechanism at each level is the same. Two neurons routinely firing together are recognized as having a common cause, as described earlier, forming synapses on a common neuron, as per below. Spatial summation is the driving mechanism.

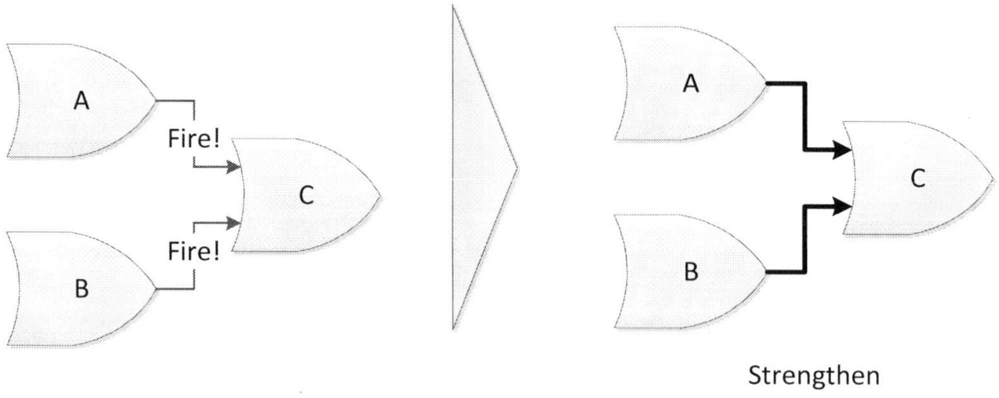

The same process occurs in hearing. Few sounds are composed of only one frequency. Identifying common pairings leads to the ability to identify a mother's voice, the timbre of a musical instrument, or the roar of a lion.

Visual Perception – Edge Detection

The example below describes how an infant learns to recognize an edge, the first step in identifying objects.

The example starts with two adjacent photoreceptors in the retina, sending their signal to two adjacent neurons to the visual cortex. Neuron A responds to light in its photoreceptive field, whereas Neuron B responds to darkness in its photoreceptive field[18].

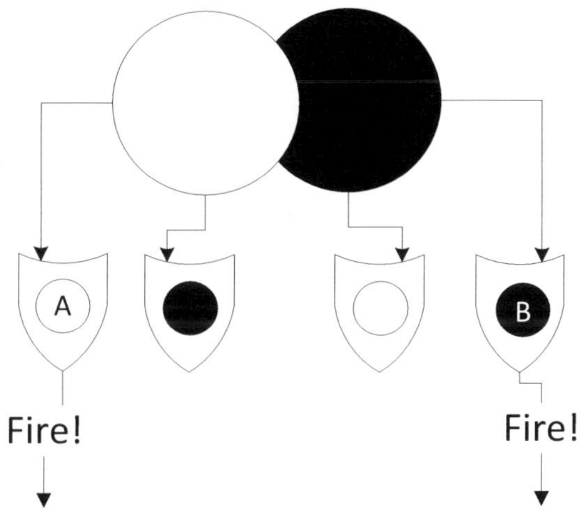

Both Neuron A and Neuron B have axons that extend to a large number of minicolumns. The concurrent firing of Neuron A and Neuron B establishes a synapse on another neuron in the same minicolumn, as pictured below.

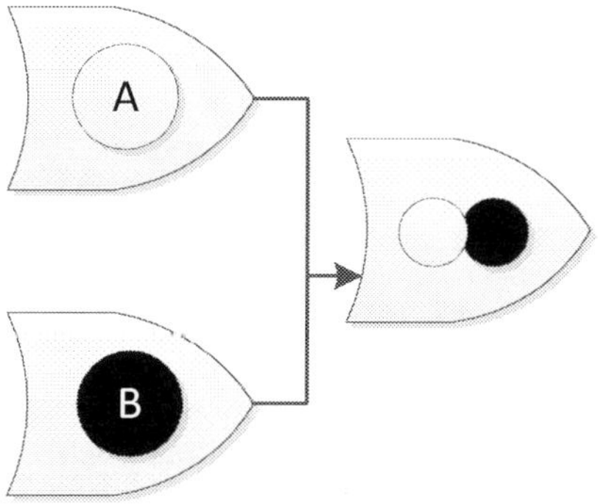

Repeated observations strengthen the synapse. Stringing a series of such detectors in a row detects an edge, as described below.

---

18 An inhibitory interneuron may be required to turn the lack of a signal – darkness – into the firing of the B neuron.

The connection process operates in a bottom-up fashion – the lowest elements tie themselves together first, then the next level, and so on. In the diagram below, two elements, whose outputs are time-correlated, connect to a **2** neuron. Then, two of the **2** neurons whose outputs are time-correlated connect to a **4** neuron.

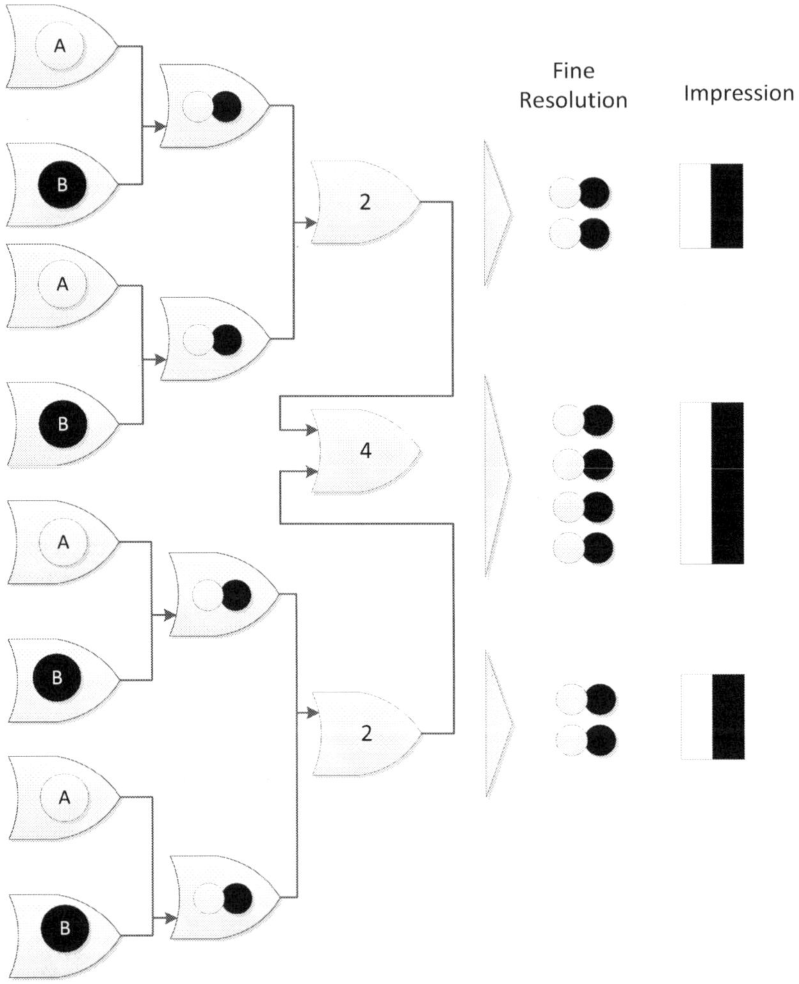

The diagram shows the recognition of the edge of a straight line. A similar scheme would recognize the edge of an arc. The visual system wires itself based on what it encounters early in life. Thus, it is biased to recognize what it has encountered previously. Connections made in early life become stronger with repeated encounters.

Ascending the Abstraction Ladder

Increasingly abstract objects are recognized as you ascend the abstraction ladder, as shown below:

Example 1 – Sequential dots become a line

Example 2 – Four lines at 90 degrees become a square

Example 3 – Four lines at slanted angles become a parallelogram

Example 4 - Two parallelograms and a square become a box

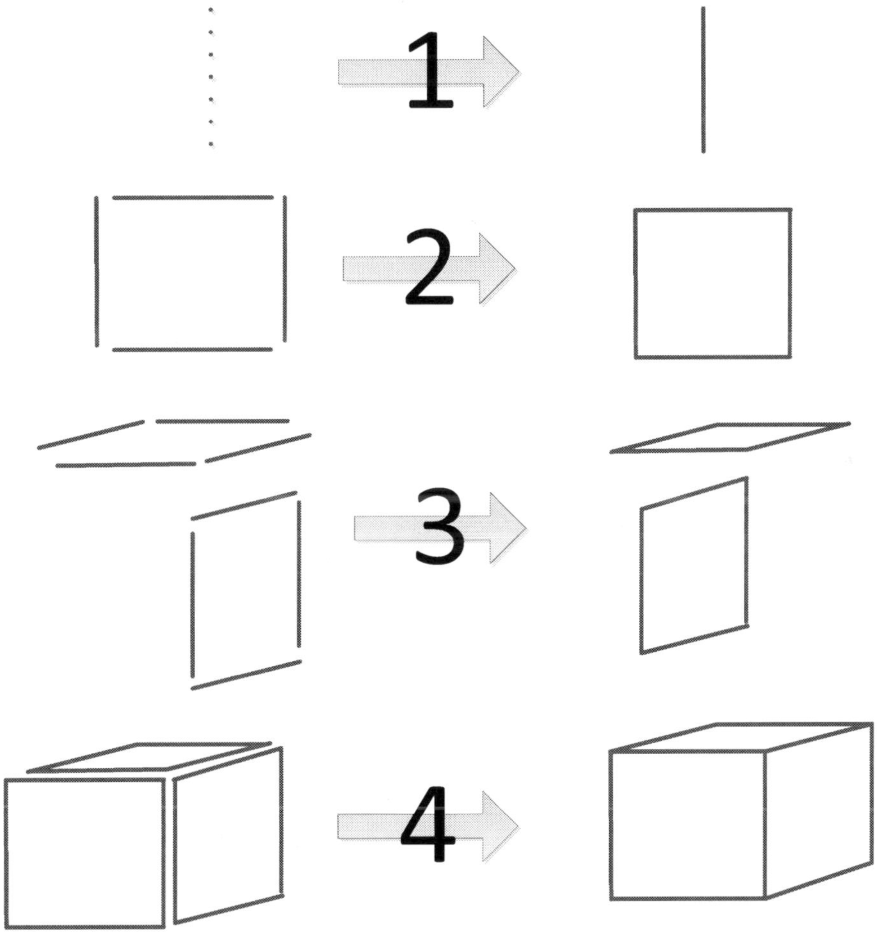

Recognizing Objects

One of the steps in the perceptual process is the recognition of objects that are more detailed than a simple cube. Gambling die is an example.

If you have never seen a die[19] before, you would observe these six faces[20].

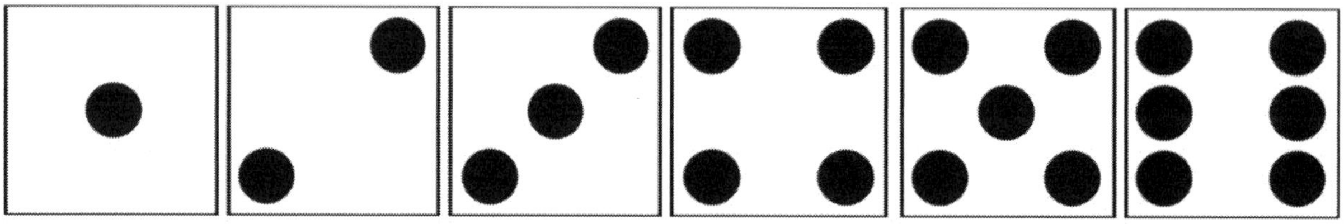

At first glance, they seem like six different things. But, when you turn it over in your hand, you discover that these are different sides of the same object, a Die. The six different symbols combine to make a single concept.

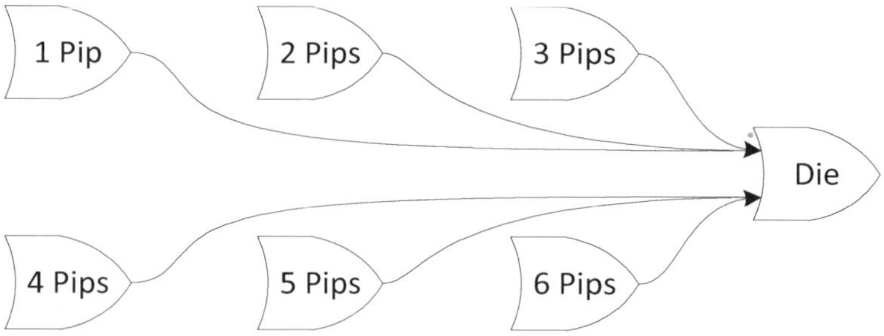

Replacing the six different symbols means you can deal with one concept – the Die – instead of six different items as you attempt to comprehend the world.

[19] Die is the singular of dice.
[20] Die_Faces.png: Nanami Kamimura; derivative work: SiPlus [CC BY-SA 3.0 (https://creativecommons.org/licenses/by-sa/3.0)], via Wikimedia Commons

Recognizing Combinations of Features - Faces

In an idealized environment, as individual faces are observed, with the order of presentation being:

1. Face Shape: Round or Oval; then
2. Eye Color: Hazel or Green

The four combinations of facial possibilities form the connections below.

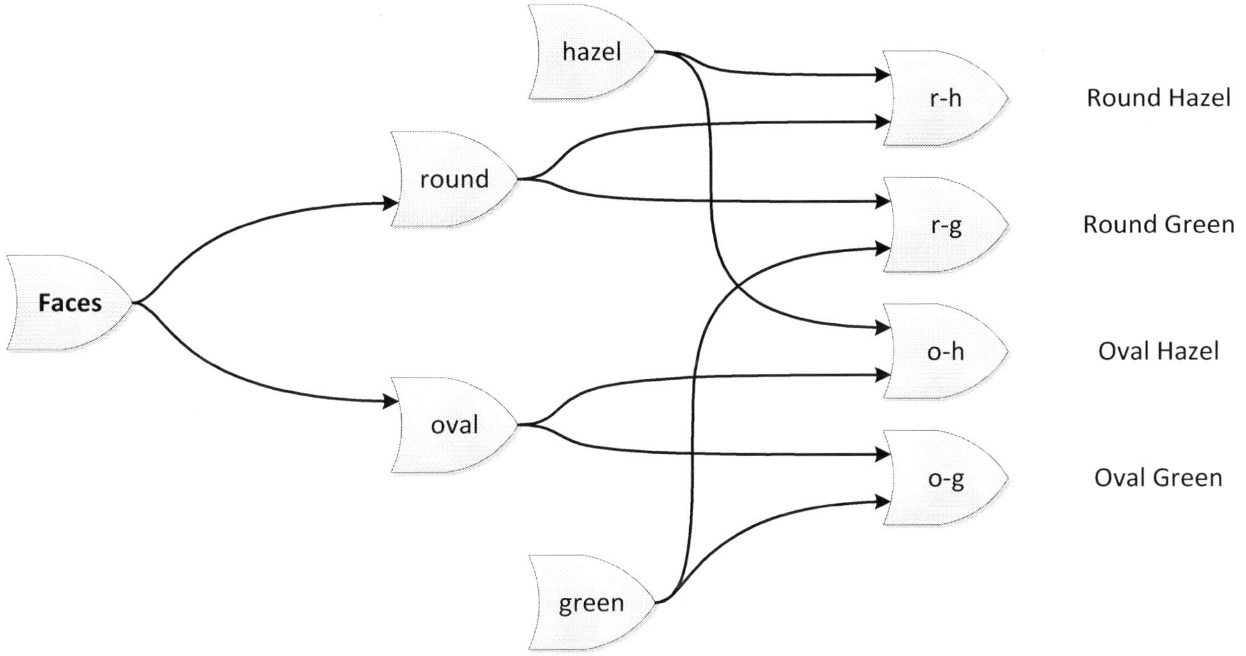

Adding blue oval eyes results in:

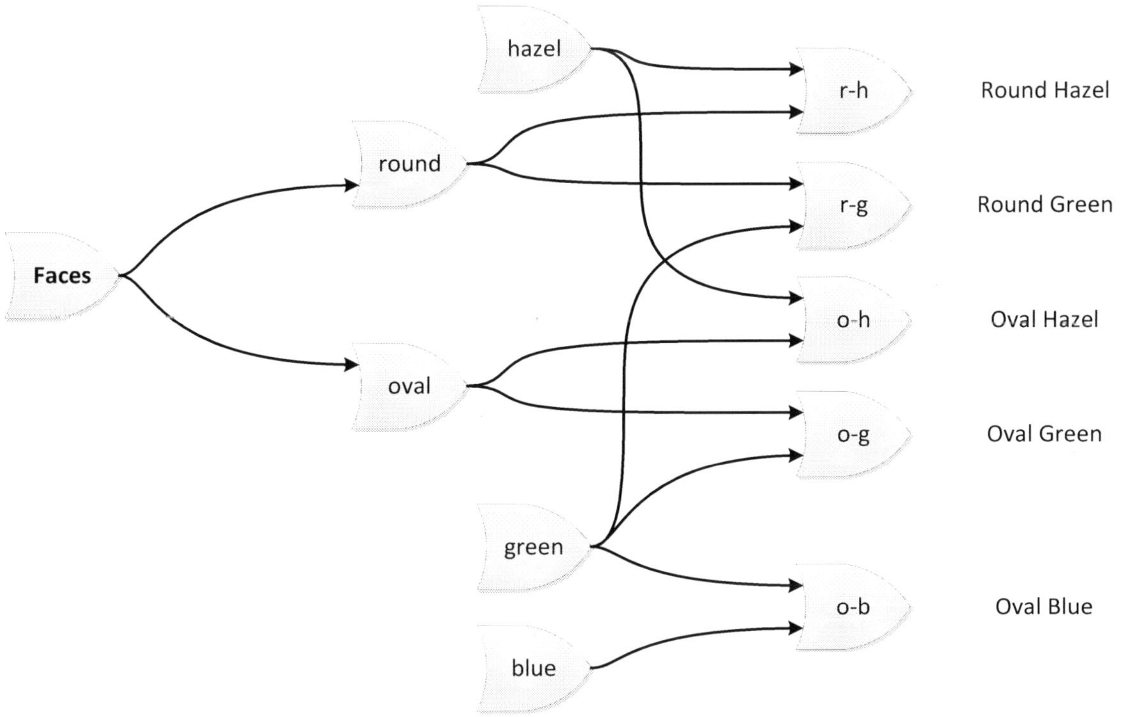

Adding blue eyes as a new distinguishing characteristic is a simple refinement. Adding a different skin color is not.

Have you ever heard a White, Asian, or Black person say?

"All you {White, Black, Asian} people look the same to me "

This politically incorrect statement reflects a deeper truth. Skin color is not a simple refinement. It is an entirely new branch, i.e.

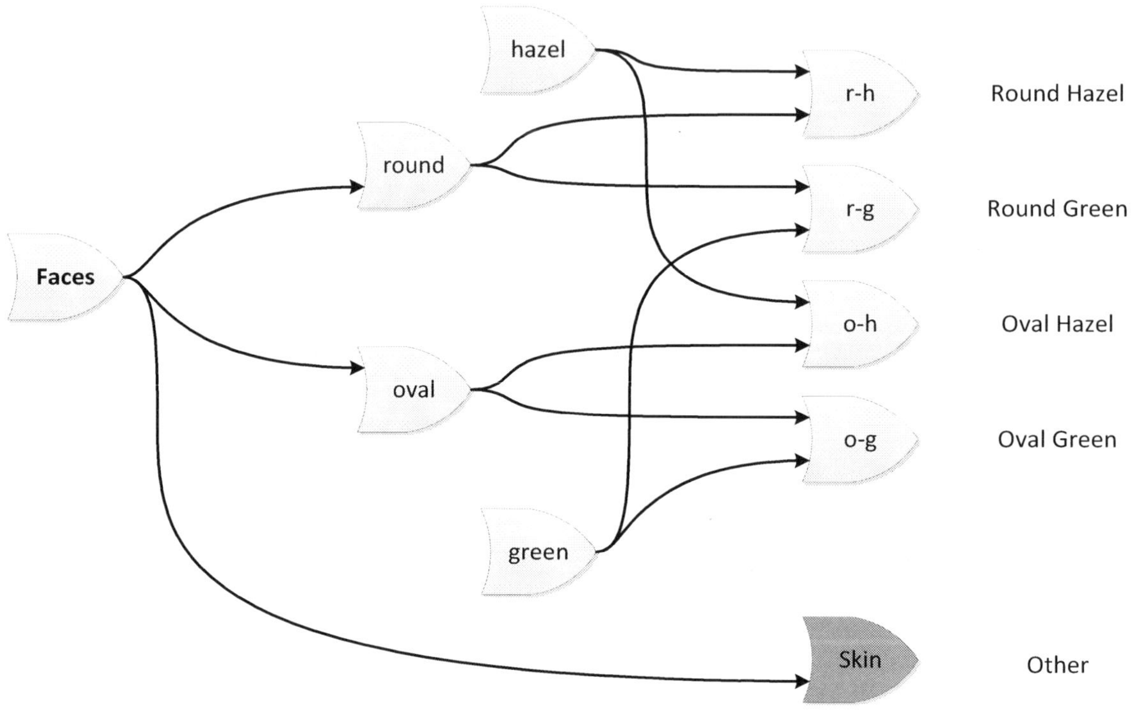

That we disagree on whether two people look alike, or not, reflects our own unique experiences and unique recognition tree.

In the Addendum, findings of research on the region of the neocortex that contains the facial recognition circuits, or Face Patch Neurons, is described.

Perception - Recognizing Movement and Location (Dorsal Stream)

Spike Timing Dependent Plasticity (STDP) was mentioned earlier as part of an argument in favor of functionally relevant synaptic plasticity rules. STDP requires that the two events occur within 50 milliseconds[21]. While that is generally too tight a time constraint, it is a good match for the mechanisms that measure movement across the visual field.

In the illustrations below, there are three neurons; A, B, and Other. Initially, the connection from **A** to Other is strong; the connection from B to Other is weak or non-existent.

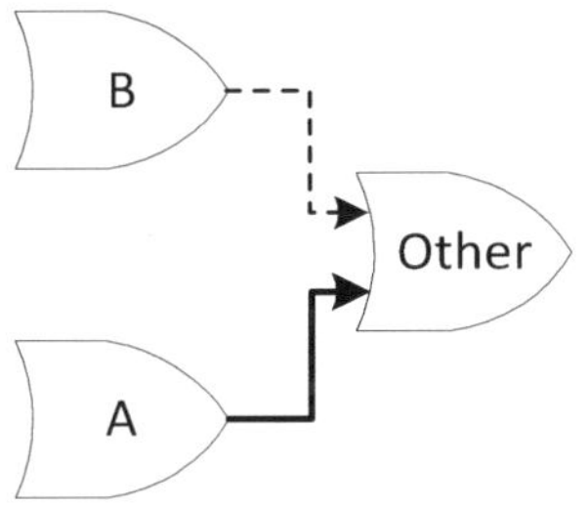

Neuron B fires, and then, within 50 milliseconds, neuron A fires.

After enough repetition, the strength of both synapses' reverses. Because B fired first, its synaptic connection strengthens. As A fired second, its synaptic connection weakens

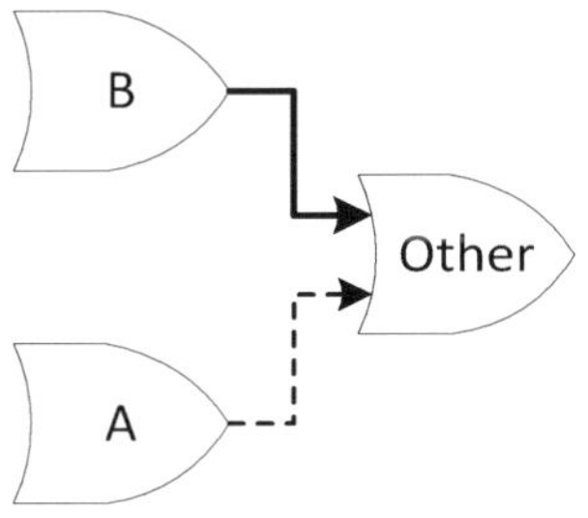

The 50-millisecond requirement of STDP is fundamental for creating synapses that measure movement across a visual field, as detailed below.

---

[21] From: A History of Spike-Timing-Dependent Plasticity - Scientific Figure on ResearchGate. Available from: https://www.researchgate.net/figure/Defining-Spike-Timing-Dependent-Plasticity-A-A-presynaptic-cell-connected-to-a_fig4_51721571 [accessed 11 Jan, 2019]

Movement Across the Visual Field

To recognize an object, you must track it – meaning that you must move your eyes/head to keep it centered in your field of view. To do that, you must quickly measure how quickly it is moving across your visual field.

In the Dorsal stream of the visual cortex, a string of neurons fire when an object is within their field of view. The rods and cones of the retina send their output to these neurons. The diagram below shows an object moving across the visual field, with each neuron in the string firing as the object passes through the neuron's field of view.

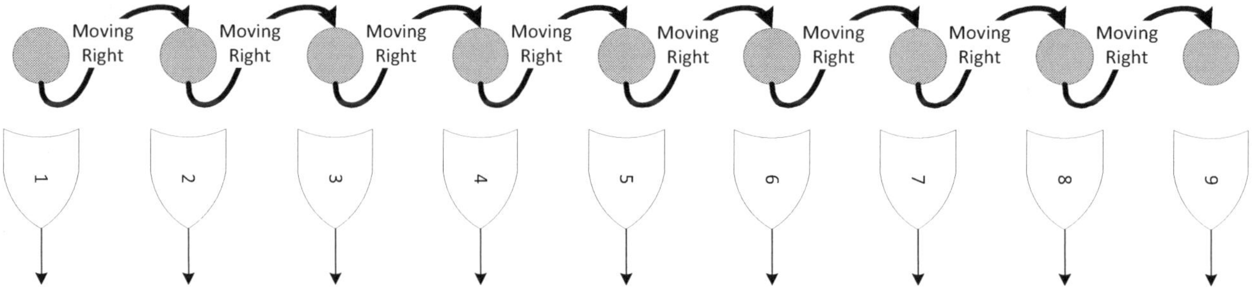

These neurons fire into a mesh of neurons. In the diagram below, the object is moving at the slowest detectable speed.

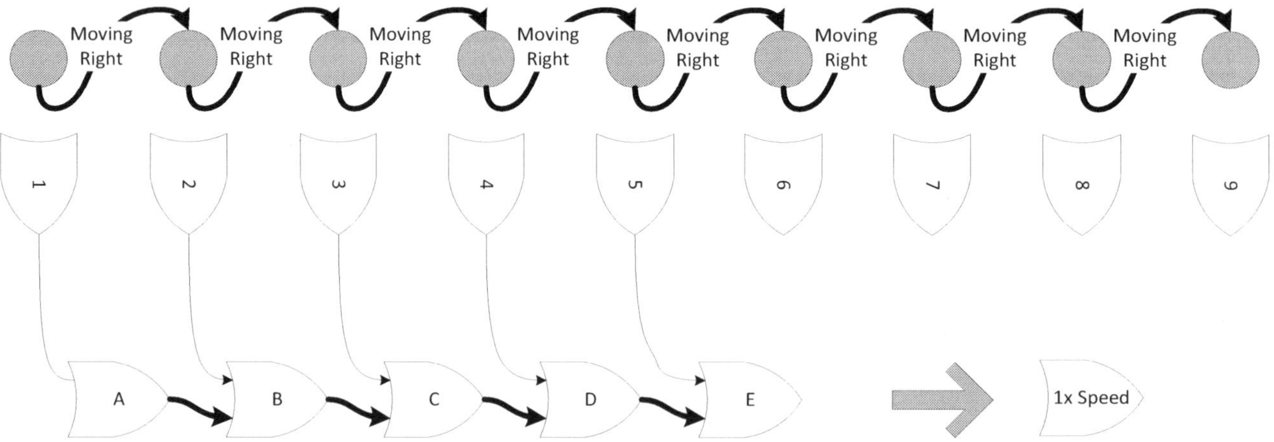

In the above, neuron A fires into neuron **B** just before neuron **2** fires., neuron B fires into neuron C just before neuron **3** fires. The firings must occur within 50 milliseconds of each other.

As per STDP, synapses will form that reflect the connections between neurons A, B, C, D, and E above.

The last neuron in the chain's firing reflects the motion of the object at the lowest detectable speed, 1x speed.

Now, double the speed of the object. Neurons A, B, C, etc. are no longer relevant. A new set of neurons intercepts the firing, L, M, and N. The following connections are created as per STDP and detect movement at twice the speed of the first example.

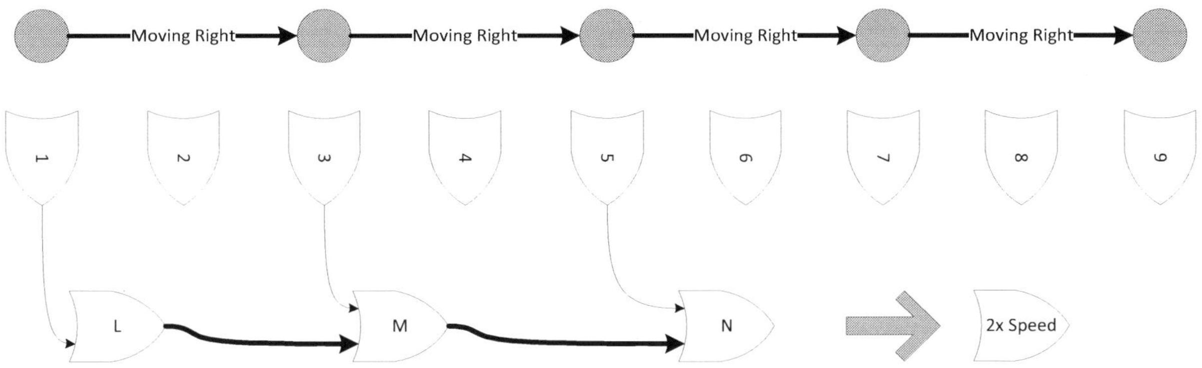

And, if the object were moving twice as fast as that, the connections established would be:

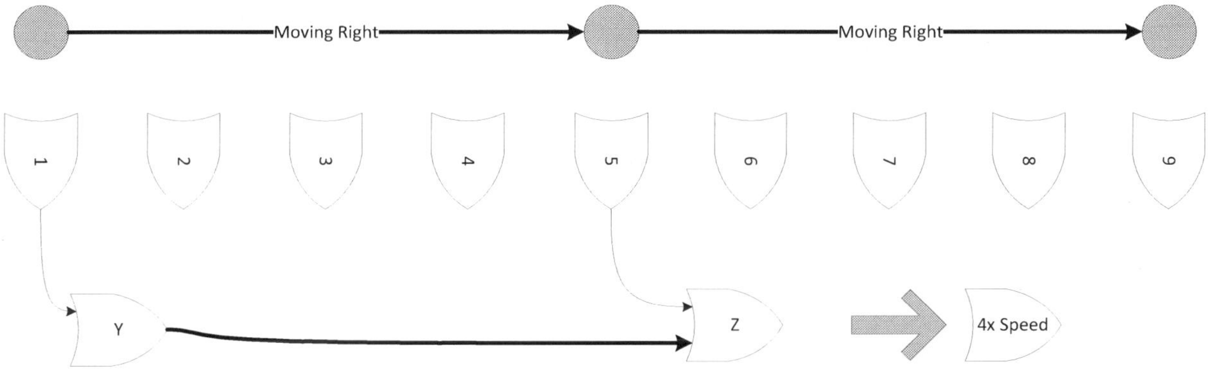

Note that detected speeds will be integer multiples of the slowest detected speed.

The speed detected will guide both your head and eye positions. After you have locked-on to the object, the detection of movement across your field of vision will tell you whether you are going too slowly or too quickly in the direction of the moving object[22].

[22] The technical term for this is Smooth pursuit. The time before lock-in is achieved is 'Open loop pursuit.' Open loop typically lasts ~100 milliseconds. Closed loop tracking is after lock-in. For more details, see: https://en.wikipedia.org/wiki/Smooth_pursuit

An Object Moving Towards You

Survival requires identifying threats that are moving towards you. Threats that are moving faster to you require a quicker response. To determine how quickly an object is moving towards you, combine edge detection with horizontal velocity measurement, as per below.

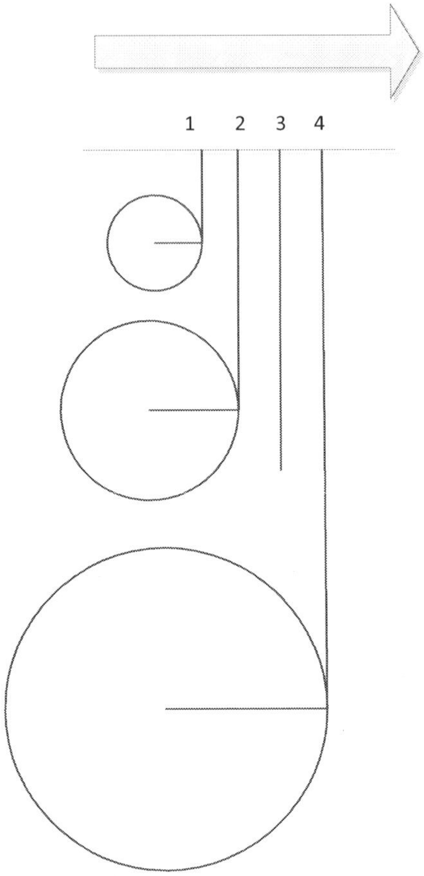

The Ventral stream contributes to detecting the edge of the object. The Dorsal stream contributes to the speed of the increase in size. The information from both streams is combined to determine how quickly an object is coming towards you.

Perceiving the Environment - Location

You are seated in a room with four walls. One wall has a window, another a door. As you turn your head in 90-degree increments, you view each side of the room. You have two sensory inputs available. One is your visual input; the other is the direction of your head. Proprioception is the name given to the sensory input that tells how far your head has turned. When you combine both sensory inputs, you create a model of the rooms, as per below.

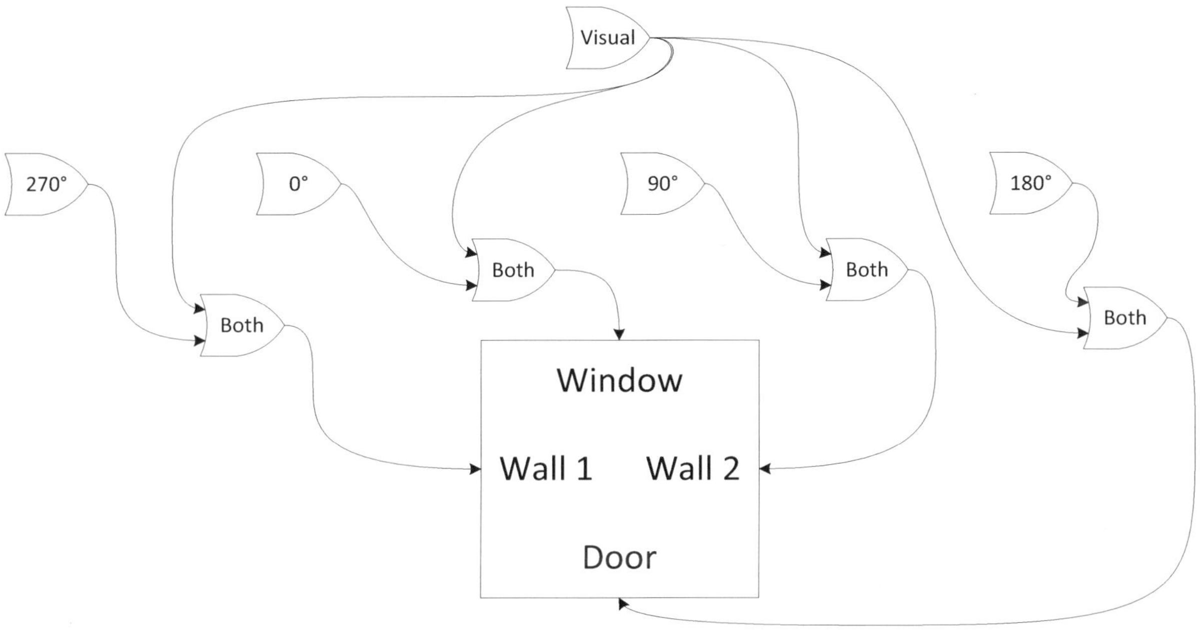

This circuitry is in the Posterior Parietal Cortex or PPC, which receives visual, auditory, and somatosensory[23] input. In the PPC, there are Grid cells that map out your current surroundings. These neurons are arranged in a hexagonal array, as shown in the picture below[24]:

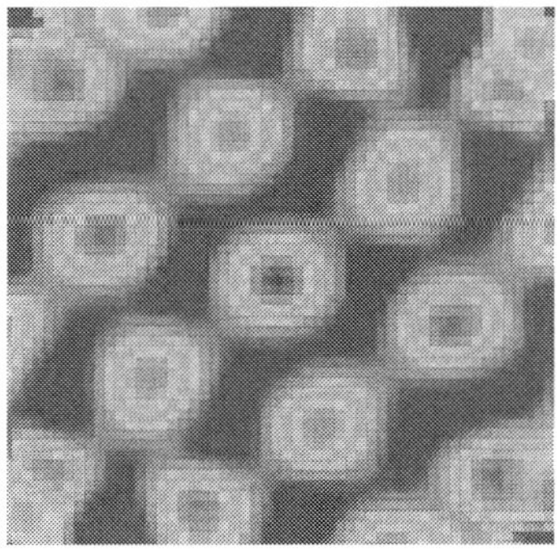

---

[23] Somatosensory input includes pressure, pain, warmth and proprioception. Proprioception is how we know where are limbs are and how they are moving.

[24] By Khardcastle - Own work, CC BY-SA 4.0, https://commons.wikimedia.org/w/index.php?curid=59507250

In light of the hexagonal array of the PPC, the increments for head movement should be 60 degrees, not 90 degrees. Next, we add another element – the extent to which the eyes point left of right, also by 60 degrees. The neural connections below show how information is combined to produce a consistent representation of the surrounding space.

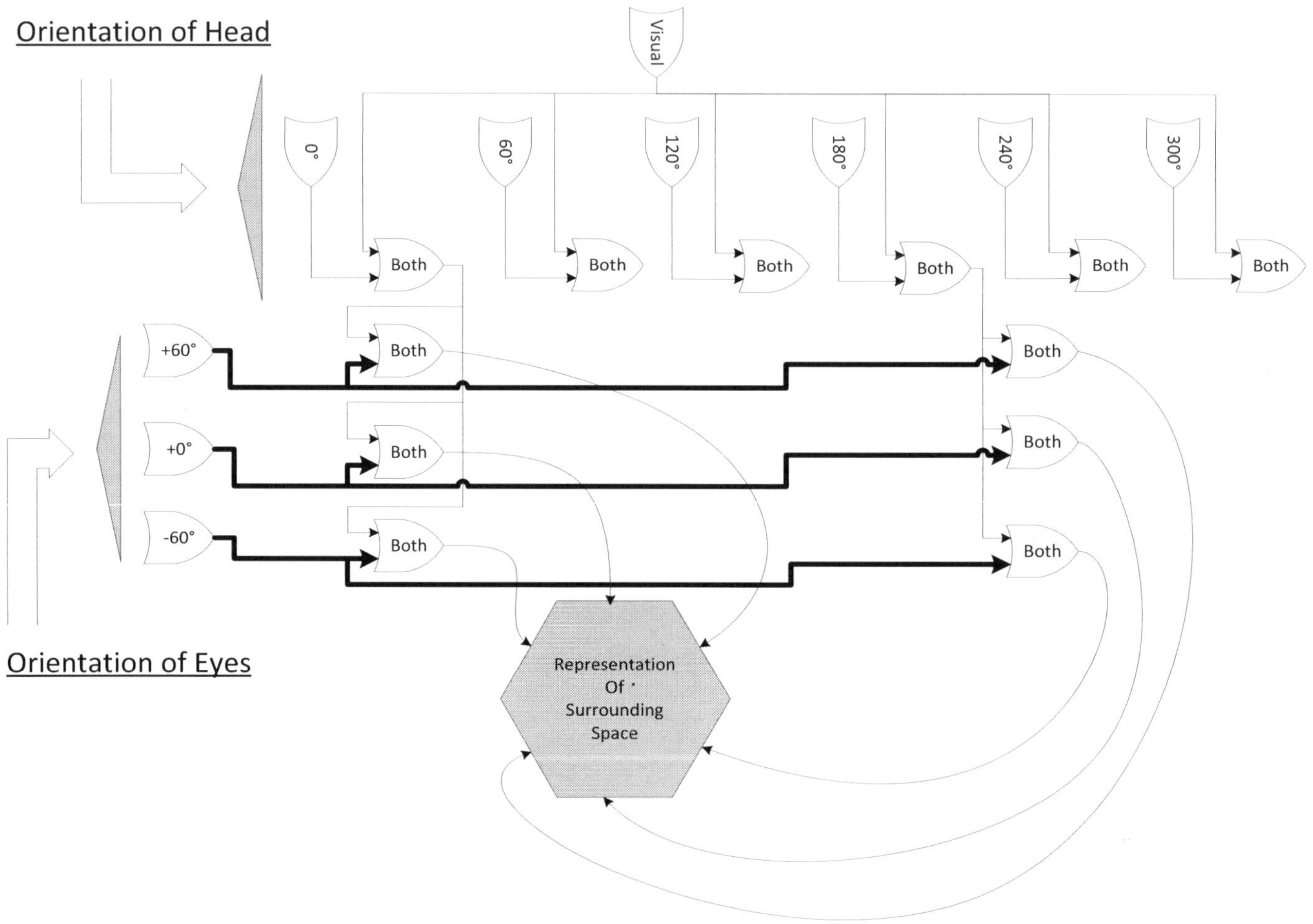

The diagram depicts the full set of 0-degree and 180-degree connections. If you wish, you can add the others.

Concept Neurons

Epilepsy is a condition where a wave of neural activity spreads over the neocortex, incapacitating the individual. When drugs, diet, or other therapies do not work, there are surgical options.

These options include cutting connections in the neocortex. Those cut connections prevent the spread of epilepsy across the entire neocortex.

The surgeon removes part of the skull. Then the surgeon uses an electric probe to test candidate areas to cut. As the patient is awake, they tell the surgeon how the electric probe affected them. The surgeon cuts across an unimportant area.

A team led by Rodrigo Quiroga[25] examined patients undergoing this surgical procedure. With the skull removed, the team implanted ~100 tiny electrodes, each thinner than a human hair into the brain of the patients. The patients viewed pictures of various people and places, such as Jennifer Aniston, Bill Clinton, and the Eiffel Tower.

Amazingly, the neurons of some patients responded to specific images. One neuron lit-up when shown the Eiffel Tower. Another neuron lit up when shown a picture of Jennifer Aniston. That neuron came to be called the Jennifer-Aniston neuron.

This observation led to the idea that a single neuron embodies a single concept. A basic premise of the Connections Model is the existence of concept neurons.

An alternative theory of neural organization is distribution representation, where a unique pattern of neural activity over a group of neurons act in the manner in which concepts are realized. Followers of the distribution representation theory use the disparaging term 'grandmother neurons' to refer to what the Connections Model calls Concept neurons.

---

[25] Invariant visual representation by single neurons in the human brain. R. Quian Quiroga, L Reddy, G. Krejman, C. Koch & I. Fried, 2005.

Concept Neuron Diagrammed

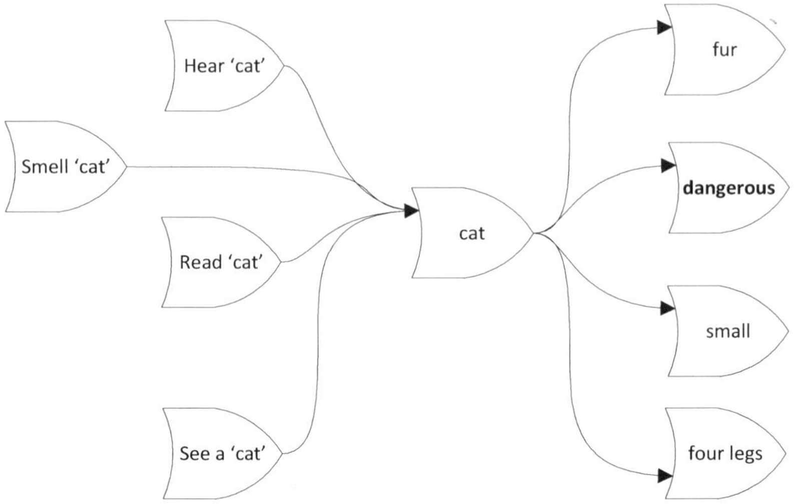

Identifiers – Lead to a Concept neuron

Properties – Follow a Concept neuron

You hear the word cat; You see a picture of a cat; You read the word cat. In all three cases, your mind arrives at the same concept – the cat (neuron) concept.

No matter which way you arrive at the cat neuron concept, you know that a cat is small, furry and has four legs[26]. Those are properties of the cat concept. When the cat axon fires, each of the property neurons also fire.

In this particular diagram, the individual believes that cats are dangerous. Perhaps that was the result of an early childhood experience. The neurons that follow the dangerous neuron dictate how the individual will react to seeing a cat.

Properties can also be 'associations.' I have a friend who, when they smell leather, is reminded of his father's leatherworking shop.

[Note that identifier neurons can be outside the neocortex. One example being smell. Smell information goes from the olfactory bulb to the thalamus to the neocortex. Interestingly, the smell is particularly effective at recalling memories. Perhaps this was a survival advantage, whether smelling carnivores or unsafe food.]

---

[26] Clearly a cat with 3 legs would still be a cat. The precise ties between one concept and its properties evolves over time to incorporate additional neural connections. Most diagrams in this description are designed to explicate the concepts. Diagrams that provided all necessary details would obscure the concept being explained.

There is an expression, "One man's ceiling is another man's floor." Likewise, properties and identifiers are not absolute but relative positions.

In the diagram below, both cat and furry are concept neurons. Cat is an identifier to furry, and furry is a property of cat.

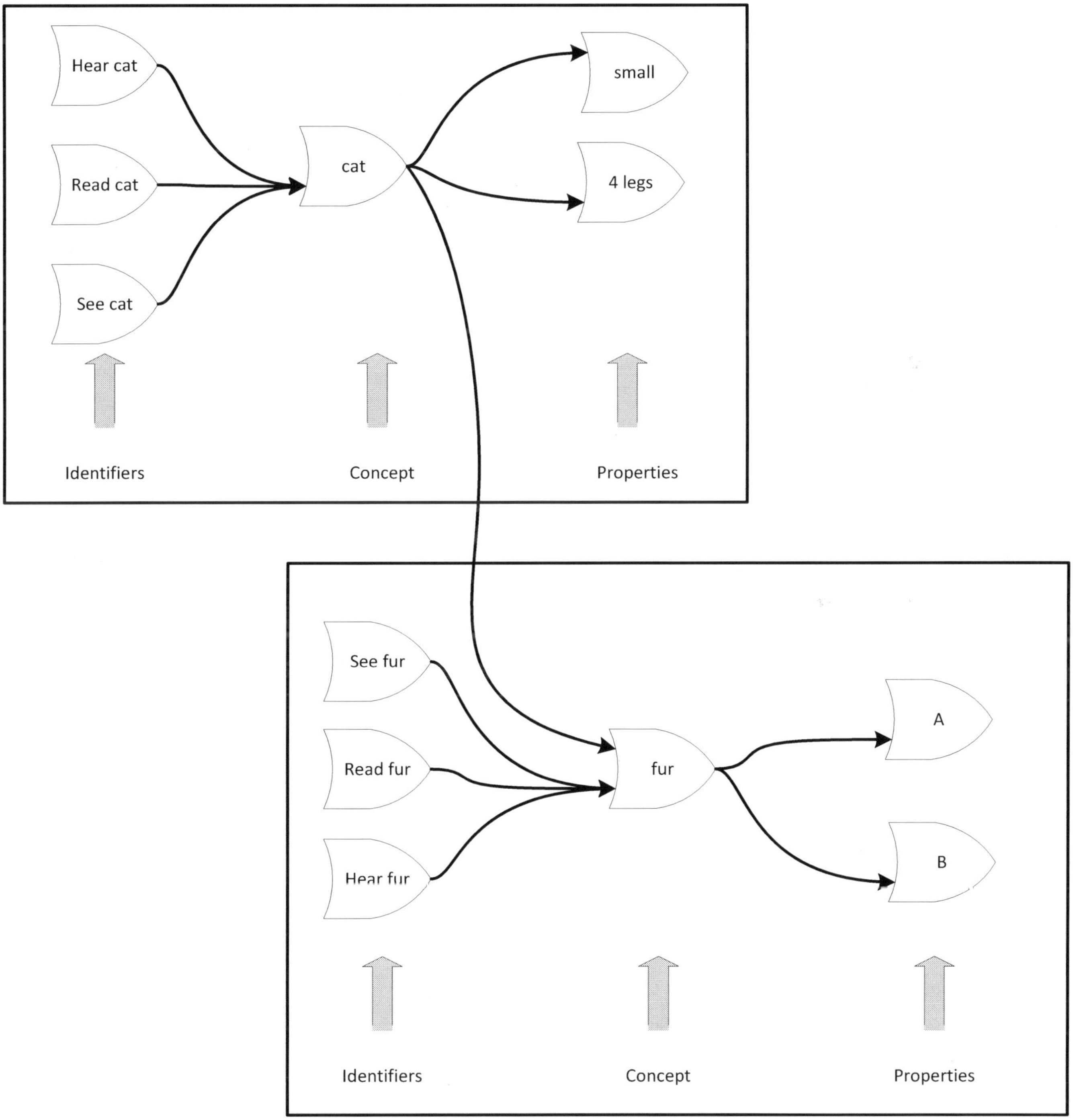

Concept Neuron Formation and Connections

You have repeatedly observed cats, and now recognize them, i.e., the synaptic connections from visual perception to the CAT concept neuron are strong.

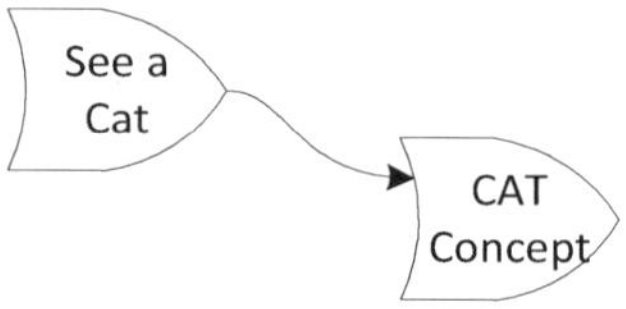

You read a book about Cats. There is a picture of a cat, with the word 'cat' beneath it. The reading of the word Cat predicts the CAT concept.

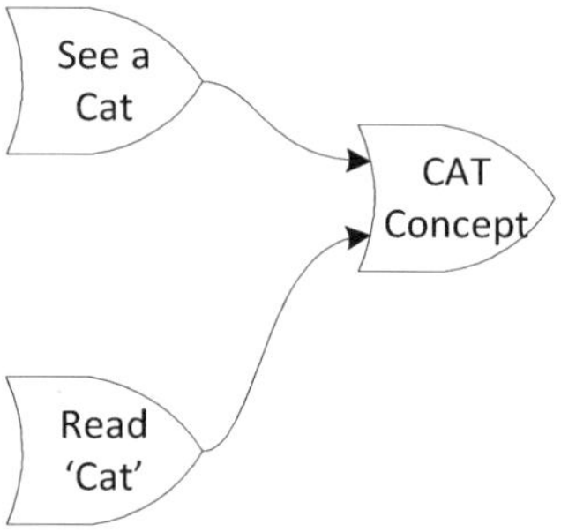

After you have learned the concept of Furry, you notice that Cats are furry.

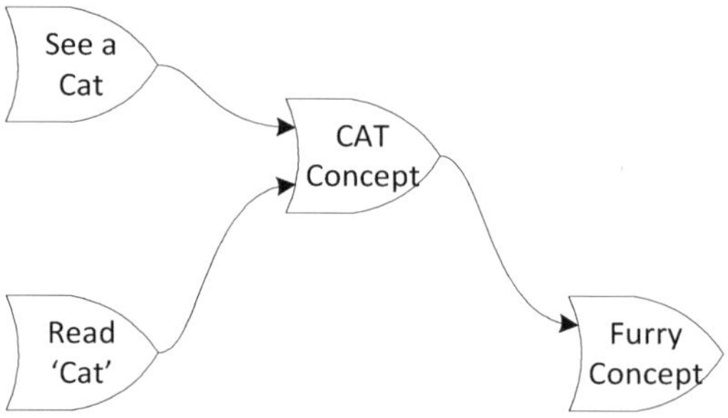

Below is a diagram that includes other related neural connections.

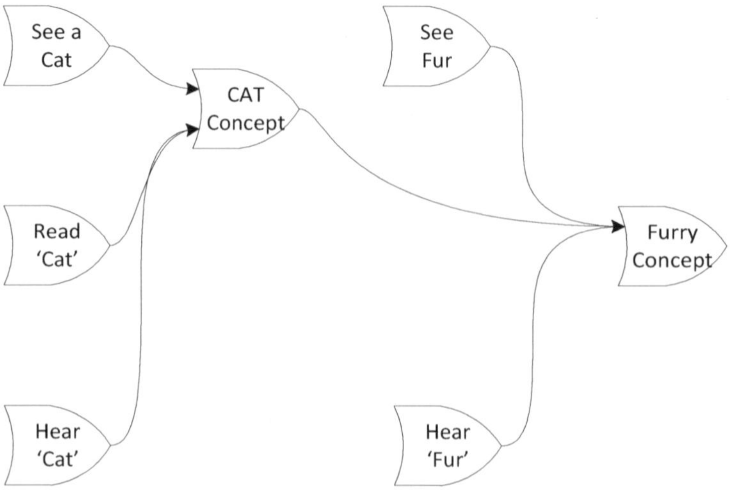

One thing leads to another

As concept neurons fire, their property neurons fire as well. Connections are both feedforward and feedback, so one thing leads to another. These connections reflect how our ideas are connected - figuratively and literally. These connections are reminiscent of a Scientific American magazine column that ended in 2001, "Connections," by James Burke. In the column, he discussed one issue, which led to a related issue, which led to a related issue, and so on until he found his way back to the original issue.

Connections from one concept neuron to another can be feedforward or feedback connections. Because feedback connections lead backward, all of the concepts that belong in a specific sub-region of the brain can fit.

In the absence of a throttling mechanism, the chain of concept neurons firing would create an avalanche of activity. Since that is not what happens, a throttling mechanism must exist. Inhibitory interneurons are likely to be involved, or the simultaneous firing would increase without limit, leading to an epileptic fit.

It is clear that when you are trying to solve a problem, the chain of related ideas can help you locate a solution. This mechanism is described further in the 'Goals' section below.

Next – Concept neurons are further characterized by whether they are:

Elementary Concept neurons; or

Category Concept neurons; or

Abstract Concept neurons

Elementary Concept Neurons

Elementary concept neurons describe something you can observe. In essence, they derive from the output of the visual cortex, or the auditory cortex, or any other sensory input. There are four types, each corresponding to a part of speech.

1. Objects that you can see – nouns
2. Properties of objects, like color, size, shape – adjectives
3. Actions, as in run, jump, move, destroy – verbs
4. Descriptive modifiers to verbs – e.g., gently, quickly, carefully – adverbs

While nouns and adjectives are observable qualities, verbs and adverbs rely on comparing time-sequential visual inputs to determine the action (verb) and its character (adverb).

The next sections detail the Elementary concept neurons that are Adjectives.

Adjectives

Categories of adjectives include:

1. What Kind – color, shape, texture, size, e.g., blue eyes, a furry cat, short woman
2. How Many – specific numbers and general qualifiers, e.g., 1, 5, multiple, few
3. Larger, Smaller, or Equal
4. How Fast – Quickly, Slowly, Faster, Slower
5. Which One – e.g., Last month, the other man

Adjectives – What Kind

Imagine that you are an infant, and the only eyes you have observed thus far are blue. You do not know that eyes can be other colors, because you have never observed the eyes of other colors. You have noticed this common occurrence and created a concept neuron that embodies the combination, for example, Blue Eyes.

Then, you observe someone with brown eyes. That leads you to create a concept neuron that embodies the combination of **Brown Eyes**.

The commonality of both **Blue Eyes** and **Brown Eyes** leads to the creation of synapses on a new concept neuron, **Eyes**. See the diagram below.

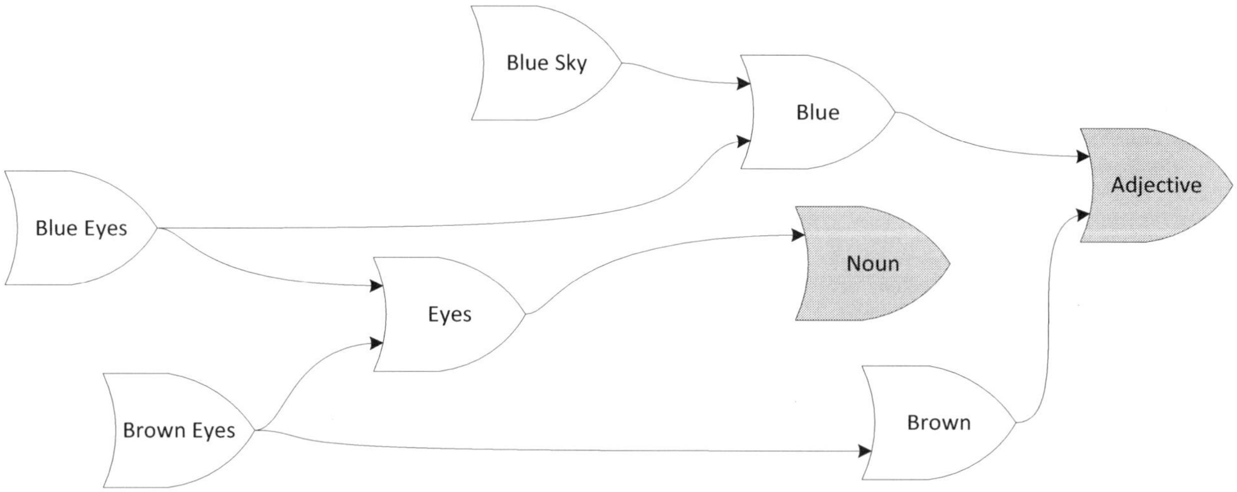

In the diagram above, there are two shaded neurons. The shaded neurons are properties that tell us whether the concept is a Noun or an Adjective. The property of a concept drives word order in a sentence.

In English, the adjective appears before the noun, as in 'Brown Eyes.' In Spanish, the noun appears before the adjective, as in ''Ojos Moradi', which translates to 'eyes brown.'

The above diagram does not include the output from the visual cortex. The diagram below includes the connections from the visual cortex (shaded) into the Brown Eyes and Blue Eyes neurons.

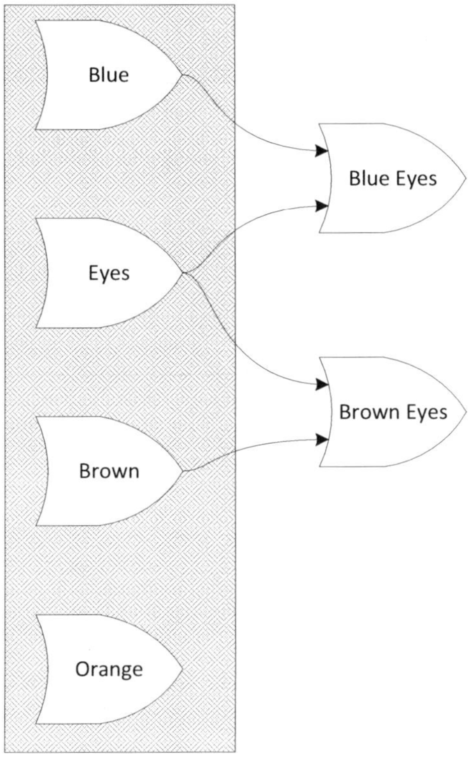

Blue, brown, orange, and eyes are all outputs of the visual cortex processing. The association cortex creates concept neurons that are representative of what it has encountered. That is why there is no Orange Eyes concept neuron.

Until the infant observed a person with brown eyes, there was no brown eyes concept neuron either. In general, the association cortex contents reflect what is known, by observation, or education.

A primary purpose of the association cortex is to guide the behavior based on predictable consequences. The properties of Concept neurons include expected consequences.

A simple example is a case where all of the blue-eyed people that pick up an infant smell bad. If a blue-eyed person attempts to pick up the infant, the infant may attempt to squirm away, to avoid the bad smell.

Adjectives – How Many

When we are young, we learn to count, i.e., 1, 2, 3, 4, etc. The diagram below shows what learning to count means.

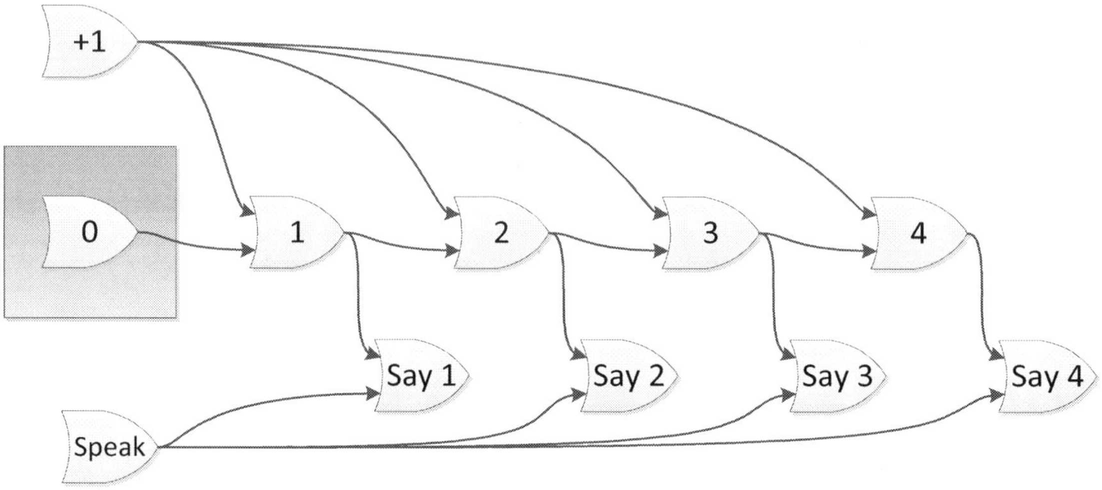

This diagram is different from the diagrams you have seen before. You do not proceed instantly from neuron 1 to neuron 2 to neuron 3. It takes some time to "Say 1". Until you have finished speaking the number 1, you can't move on to neuron 2. For now, accept there is a way to hold on to where you are in a list[27].

You are looking at newly baked cookies. You are to divide them fairly with your siblings. You need to count them. Counting the cookies requires that your attention moves from one cookie to the next. Each time you land on a cookie, you increase your count. As you count each cookie, you move down the list of numbers, one at a time.

The numbers are not just abstract points in a sequence; in your memory, there is an image that illustrates the meaning of each number, i.e.

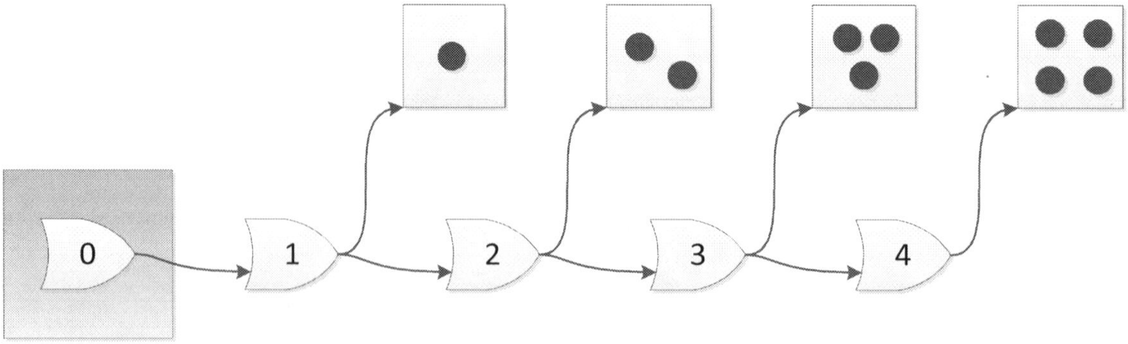

[27] The ability to hold your place in a sequence is a major theme of the next chapter.

Adjectives – Larger or Smaller

In school, we memorize addition and multiplication tables of single digits. There is another table that we unconsciously learn over time. We use it when deciding which of two numbers, with the same quantity of digits, is larger or smaller[28].

	0	1	2	3	4	5	6	7	8	9
0	=	<	<	<	<	<	<	<	<	<
1	>	=	<	<	<	<	<	<	<	<
2	>	>	=	<	<	<	<	<	<	<
3	>	>	>	=	<	<	<	<	<	<
4	>	>	>	>	=	<	<	<	<	<
5	>	>	>	>	>	=	<	<	<	<
6	>	>	>	>	>	>	=	<	<	<
7	>	>	>	>	>	>	>	=	<	<
8	>	>	>	>	>	>	>	>	=	<
9	>	>	>	>	>	>	>	>	>	=

When determining which number is larger, or smaller, we start at the leftmost digit and look up the pair in the table above. When the two digits are not equal, the table tells us which number is larger.

When the two digits are the same, we proceed to the next pair of digits and repeat. If all digits are the same, the two numbers are equal.

Concept neurons encode and store the addition, multiplication, and the table above. You can imagine that the combination of the two digits – across and down – select a concept neuron whose property is either larger, smaller, or equal.

Note that procedures often require keeping track of intermediate results. When the amount of information is more than we can keep in our head, we resort to recording the results on paper or in computer files.

---

[28] Unsigned or positive numbers.

Category Concept Neurons

In the diagram below, we see that carrot, celery, and kale are all members of the vegetable category and that apple, pear, and grape are members of the fruit category.

All of the members of a category share common properties. In the example below, we see that both vegetables and fruits are members of the plant category. There is no need to attach the plant property to the carrot concept neuron. When the carrot concept neuron fires, the chain of connections will fire the plant property as well, i.e.

carrot => vegetable => plant

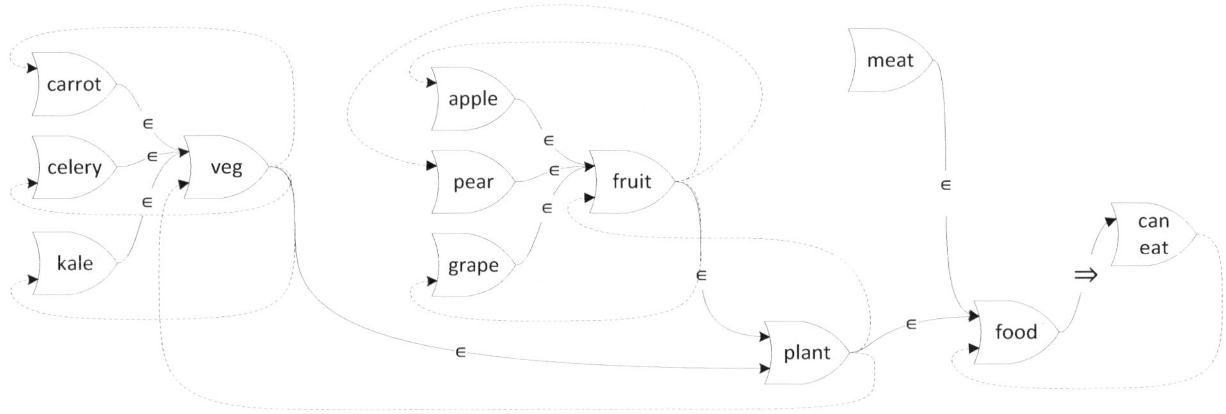

The dashed connections above are optional feedback connections. They are not always present. A feedback connection means that you will have no trouble naming an example of a vegetable. Without the feedback connection, you would need to search far and wide for an example concept neuron that leads to the vegetable concept neuron. That might take some time.

With the dashed connection, you can immediately go from vegetable to carrot, celery, and kale.

Abstract Concept Neurons – Procedurally verified

An Elementary concept neuron represents a concept closely tied to sensory output, i.e., observed.

Category concept neurons can be enumerated.

Abstract concept neurons are not observable or enumerable. A procedure, or set of steps, is needed to verify that the concept is valid for the currently identified elements.

One example of an Abstract concept neuron is the notion of Justice.

But what is Justice? The definition isn't particularly helpful, but a list of synonyms helps.

Fair Play, Impartiality, Lack of Bias, Neutrality, Objectivity

Justice means all of these things, and that they are simultaneously satisfied. Each of the synonyms is a property concept neuron of the concept neuron Justice.

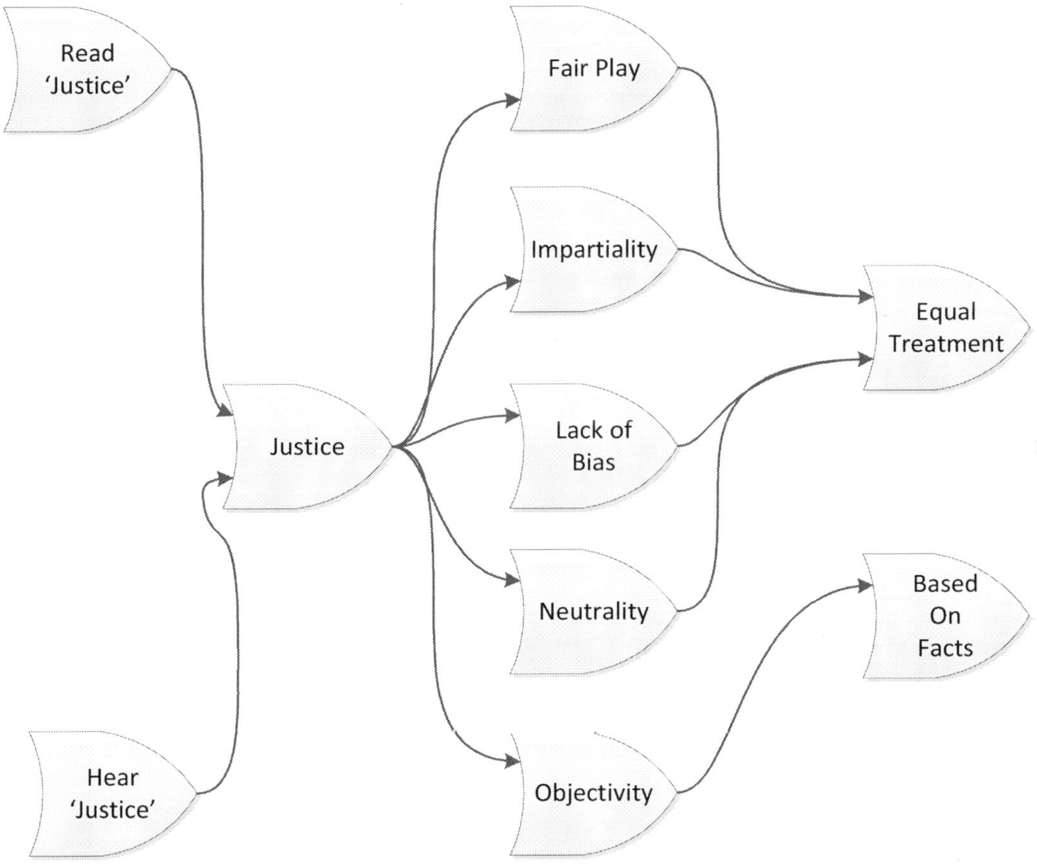

To determine if Justice is present, we verify that all of the properties are satisfied. As per above, equal treatment is a major common property. The particular circumstances dictate which of the four identifiers to Equal Treatment are most applicable. The procedure below is highly speculative.

First, we must be able to identify/enumerate all of the groups to be compared. In the example below, that is accomplished via feedback connections from the Ethnic group's category concept neuron to each element of the category.

Second, we compare the treated properties, both treated well and treated poorly. The solid connections show groups A and B treated well, but groups C and D treated poorly. A mismatch of treatments indicates unequal treatment.

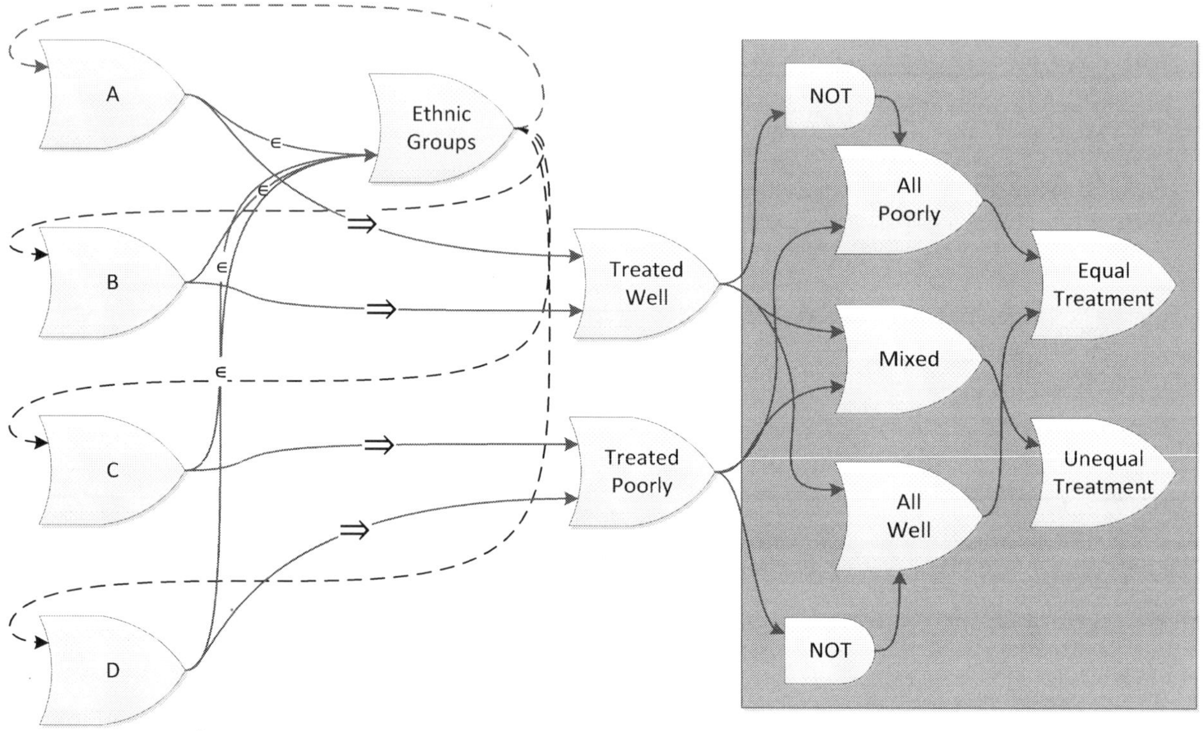

The wiring on the right implements the logic in the table below:

Treated Well	Treated Poorly	Conclusion	As
Any	None	Equal Treatment	All Treated Well
None	Any	Equal Treatment	All Treated Poorly
Any	Any	Unequal Treatment	Treated Unequally

The connections on the left, are ordinary category concept neuron examples whose ethnic group property determines whether they are treated well, or treated poorly. The connections in the shaded box could be created based on a correlation of each of the combinations of treated well, NOT (treated well), treated poorly, and NOT (treated poorly) resulted in a concept neuron of the combination.

Testing a Hypothesis

The above scheme is an example of testing whether a hypothesis is correct. Another way, illustrated below, relies on identifying an outcome or prediction that is at odds with the facts. Neither method is foolproof, as you may not have thought of all possibilities. Sometimes, the number of elements in a group makes exhaustive testing all elements or combinations impractical.

Two Opposing Ideas at The Same Time

Once I was with a friend when we both noticed a layer of fog above the snow. It was a warm day, and the snow was disappearing quickly. The friend expressed how quickly fog could eat away snow.

I said that it was the opposite - that it is the rapid sublimation of snow is exceeding the capacity of the air to hold the moisture, thus creating the fog.

My friend was not convinced. We each had two opposing beliefs, or ideas, i.e.

Idea #1 = [Fog] Leads To [Accelerated Elimination of Snow]

Idea #2 = [Accelerated Elimination of Snow] Leads to [Fog]

This circumstance reminds me of a favorite quote from F. Scott Fitzgerald is:

> "The test of a first-rate intelligence is the ability to hold two opposed ideas in mind at the same time and still retain the ability to function."

How might these two contradictory ideas be represented in neurological structures? First, consider them individually.

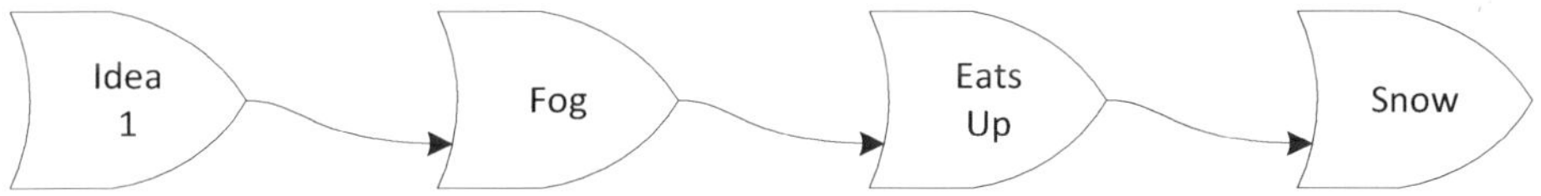

And

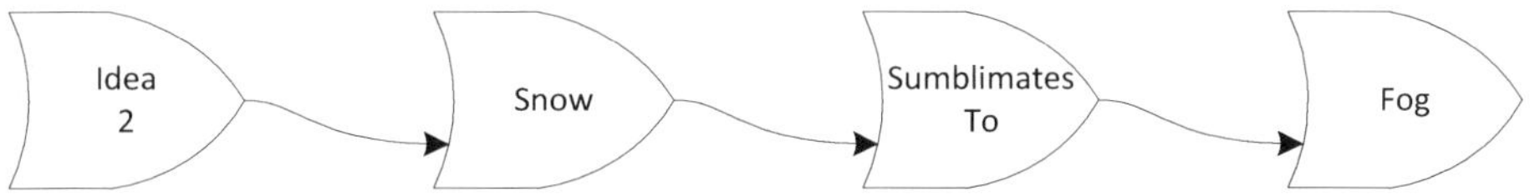

My friend was sure of Idea 1, and I was sure of Idea 2. How to resolve this impasse?

Following F Scott Fitzgerald's advice, we need to hold both ideas in our head at the same time.

One scheme for holding both would be to make both the hypothetical fog and the hypothetical snow and look for a contradiction with known facts.

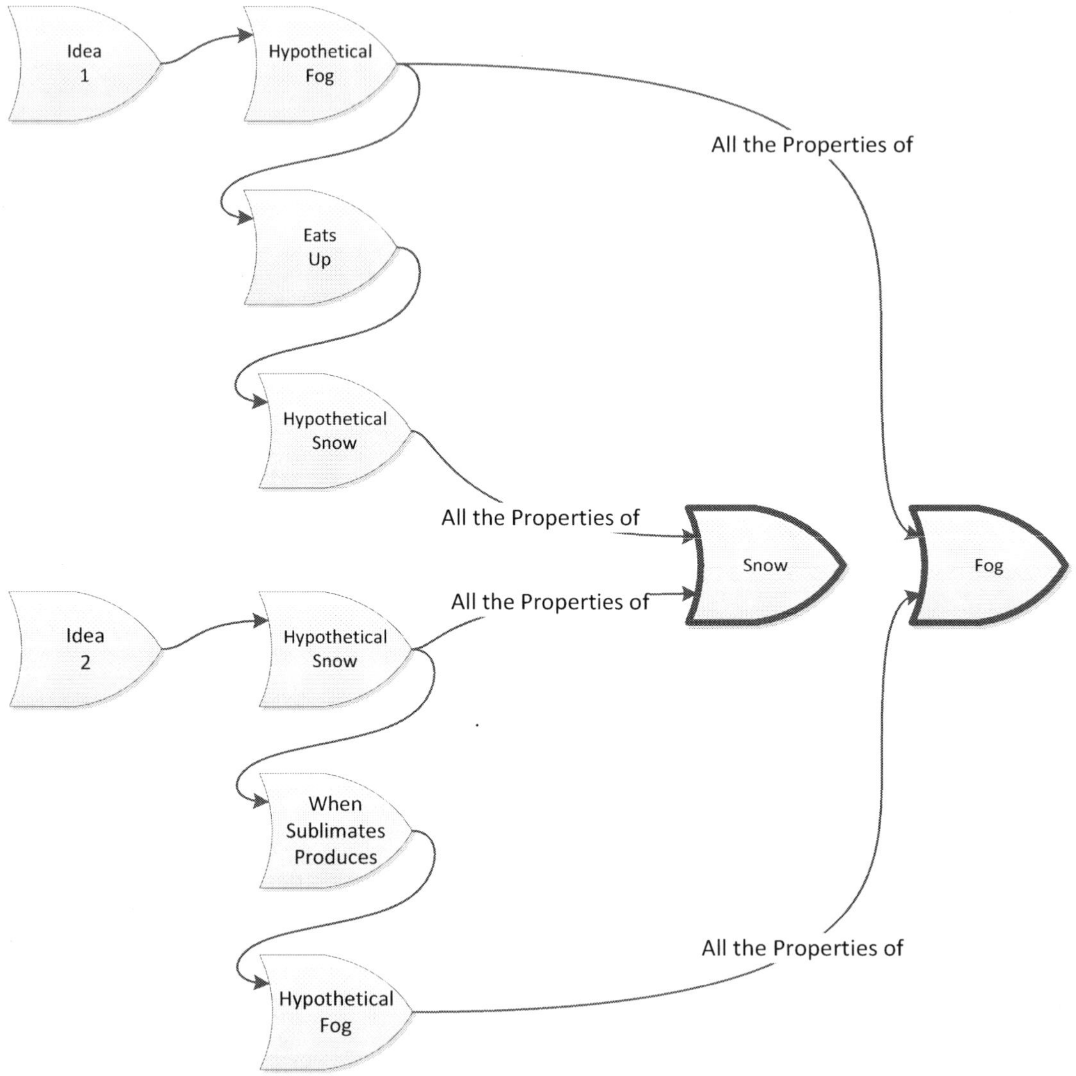

These are the critical facts:

#1 – Snow evaporates directly into the air (sublimates) – without entering a liquid state

#2 – Air with more moisture than it can handle (supersaturation) is fog.

These facts are represented below in shaded neurons.

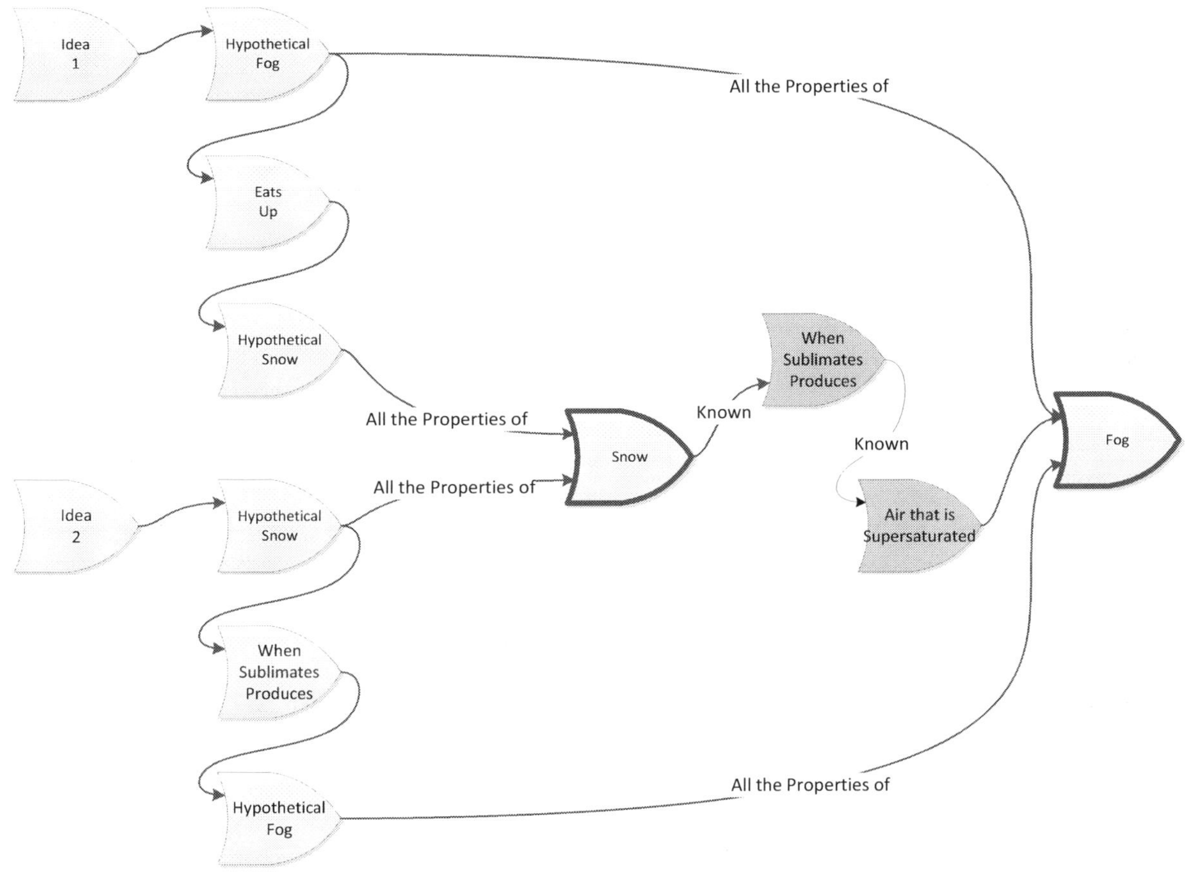

There is a path from Hypothetical Snow to Hypothetical Fog (idea 2), but not from Hypothetical Fog to Hypothetical Snow (idea 1). Idea 2 is consistent with the facts.

Often, people evaluate what they hear against what they believe is true. If they hear something inconsistent with their beliefs, they will reject it. To do otherwise requires the ability to consider both their current beliefs and an alternative belief. Only a person with a first-rate intelligence can do that.

Additional Concept Neuron Connection Diagrams

The diagram incorporates the notion that a three-legged cat is still a cat. Perhaps the number of legs is not an essential identifier, after all.

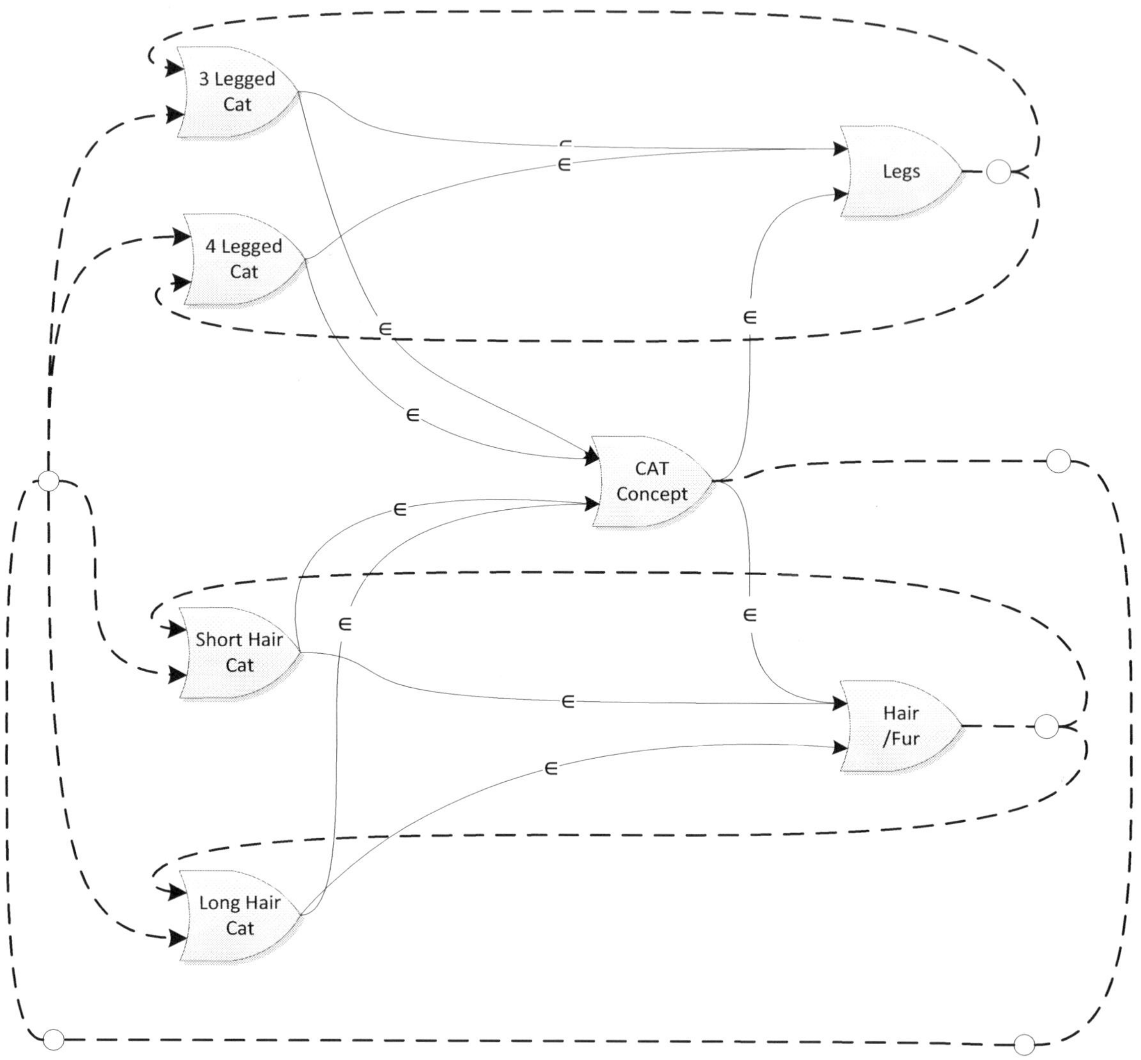

And this more extensive diagram on the following page.

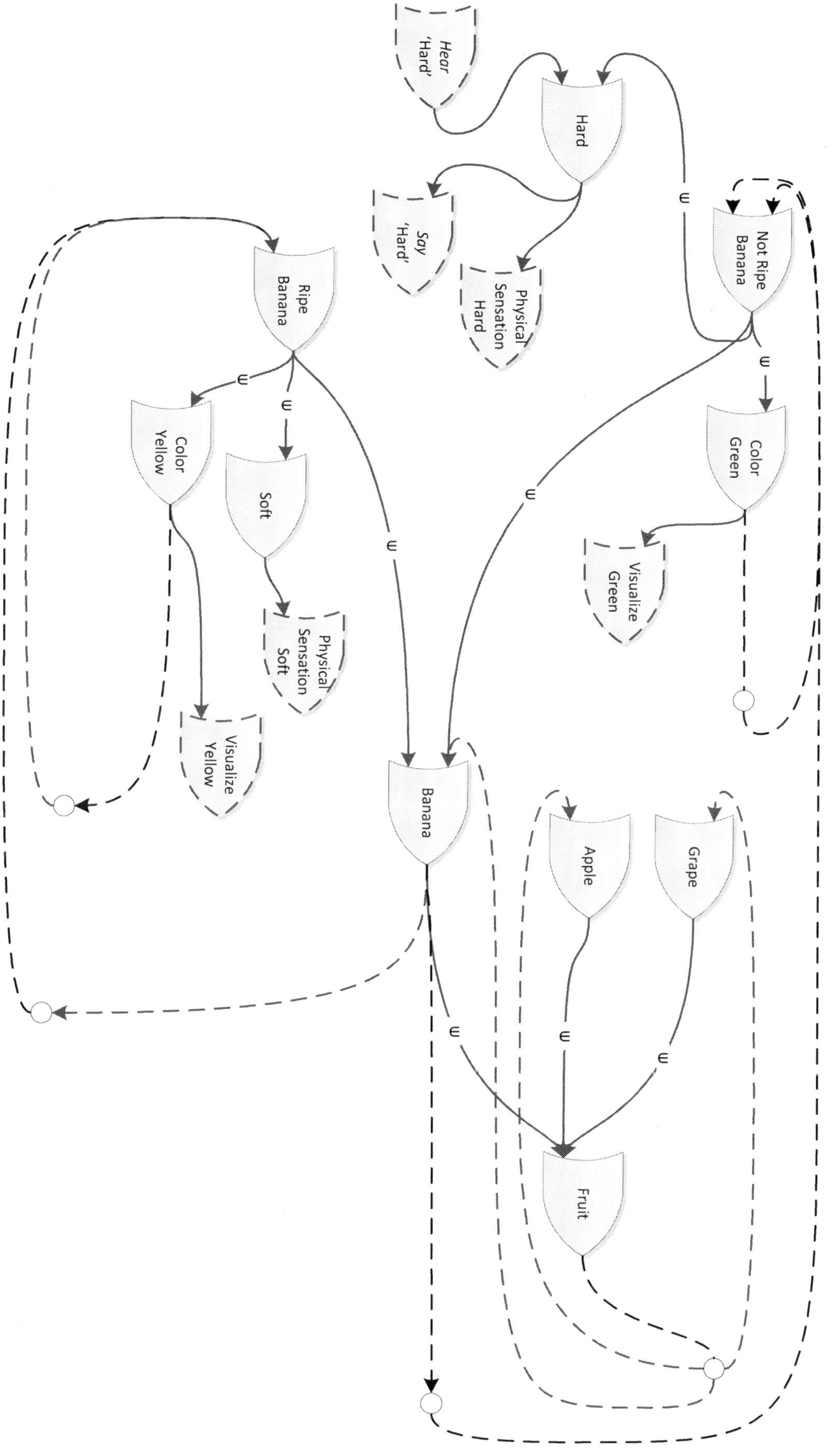
Hear 'Hard'
Hard
Say 'Hard'
Physical Sensation Hard
Not Ripe Banana
Ripe Banana
Color Yellow
Soft
Color Green
Visualize Green
Physical Sensation Soft
Visualize Yellow
Banana
Apple
Grape
Fruit

Motivation

This section is not focused on the Connections Model per se but does provide context on how it fits in with other behaviorally relevant elements.

Pain

Physical pain comes from nerve endings that detect pending or actual damage. Pain sensors are in the skin and internal organs. We avoid pain, but would we be better without it?

Jo Cameron, a Scottish woman,[29] has never felt pain. She did not feel pain during childbirth, nor surgery on her hand. The price of not feeling pain is that she burns, or cuts herself, quite often. Experiencing pain appears to be a requirement for learning to avoid it.

Pain is how we learn that we should not be doing something. We avoid actions that have, or would likely, lead to pain. Those actions include ones that lead to injury or death. Thus, feeling pain, and avoiding it, improves our chances of survival.

Mature Behavior

Cortical interneurons are responsible for inhibiting neural activity. When our prefrontal cortex recognizes that a particular action will lead to pain, a cortical interneuron will fire, preventing what would otherwise be a painful result.

The prefrontal cortex, being at the far end of the abstraction hierarchy, is the last portion of the neocortex to mature. For most people, it does not mature until the late teens or early twenties. Without the wisdom of the prefrontal cortex, young people are more likely to do things they will later regret.

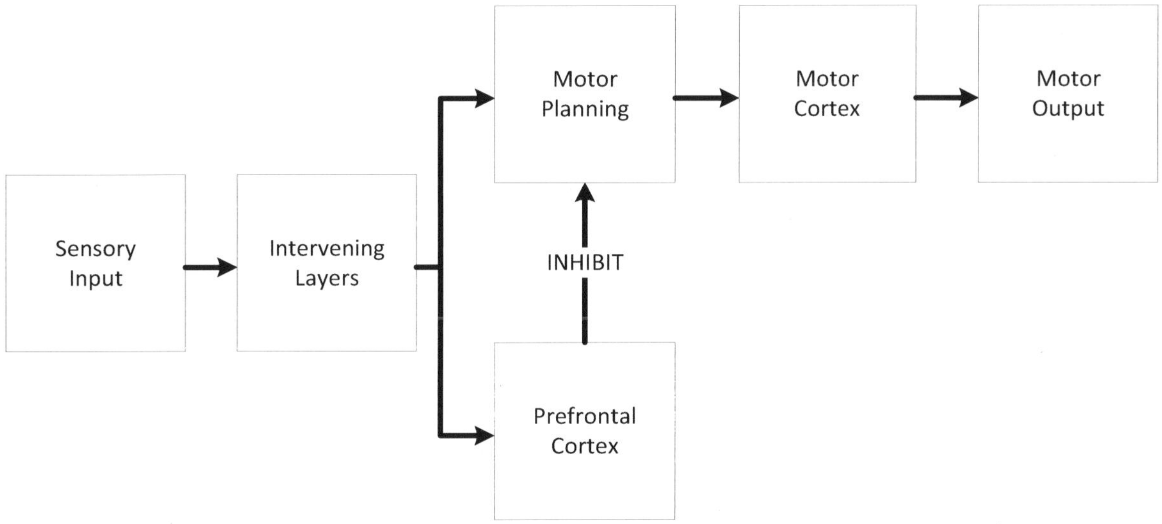

[29] The NY Times Article of March 26, 2019 titled "At 71, She's Never Felt Pain or Anxiety. Now Scientists Know Why". Additionally, she heals more quickly than we do. Could pain interfere with the healing process?

Operant Conditioning

Operant conditioning is a procedure in which animals learn when to expect pleasure (e.g., food), or pain (e.g., electrical shock.) Thorndike generalized his finding in his law of effect, which states[30]:

> "Behaviors followed by satisfying consequences tend to be repeated, and those that produce unpleasant consequences are less likely to be repeated."

Operant conditioning works with many animals, including those without a neocortex. Thus, the mechanisms implementing Operant Conditioning are likely outside the neocortex. It turns out that both pain and pleasure result in the release of dopamine. Dopamine facilitates the creation or strengthening of synapses that encapsulate the prediction of what is to come.

An even more fundamental component of motivations is the primary drives that are essential for survival and reproduction, described below.

The Four F's of behavioral Psychology

Students of neuropsychology learn that there are four basic drives or instincts of animal behavior.

1. Fighting
2. Fleeing
3. Feeding
4. Procreation[31]

These behaviors are essential to the survival of the species. They exist in animals without a cerebral cortex. They are wired in, not learned. Current thinking is the activity in the Hypothalamus, which is not part of the Cerebral cortex, drives these behaviors.

Nonetheless, I propose that animals with a neocortex (mammals) have a fifth F.

---

[30] https://en.wikipedia.org/wiki/Operant_conditioning

[31] Substitute another word that begins with F.

Learning is Fun

All mammals, with their expansive neocortex, are capable of learning. While an animal can learn from a deadly encounter, when they survive it, it is so much safer to learn while playing.

Playing is how mammals learn without being engaged in dangerous situations. During playtime, mammals learn how the world works. By play fighting with their siblings, they learn how to fight. By running away, they learn how to flee.

When mammals are learning, the brain secretes dopamine. Dopamine reinforces the joy of the experience and helps in the creation of new synapses. New synapses encode learning about actions and consequences.

Because Learning is enjoyable, we tend to do it. The more we do it, the more we learn. The more we learn, the better our chances of survival when the circumstances are life-threatening.

For students to learn effectively, they must be having fun. If they are not having fun, if they are struggling, and they are not learning, they are in pain. Both painful and pleasurable experiences result in the secretion of dopamine. With pleasure, the lesson learned is the new information. With pain, the lesson learned is how to avoid learning.

There is much research on the best ways to teach, yet, when students are not having fun, the methods are irrelevant. In that education builds upon earlier teaching, failure to understand and learn at any point leads to greater difficulty in learning new material. Failure begets failure.

Learning Facts

The diagram below shows the connections after a student learns that Columbus sailed in 1492. The bold items are the newly acquired concepts and connections.

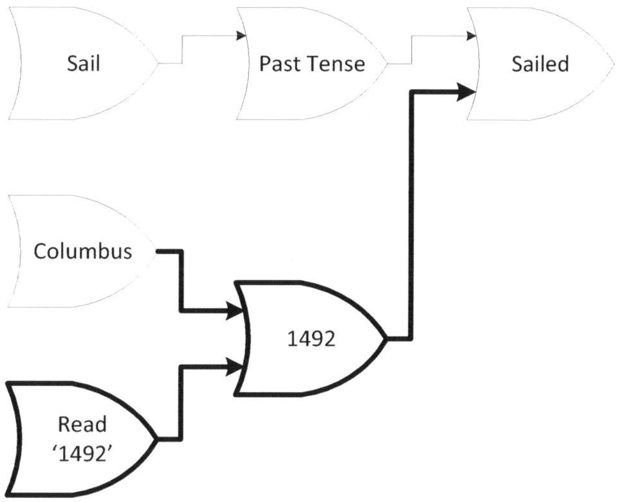

In the next section, we discuss how neural activity operates at behaviorally relevant times scales, explaining how we hold onto a thought or remember where we are in a sequence of actions.

Hold that Thought

Focus

A neuron loses its elevated state in milliseconds. Real-world events span much longer periods. How is it that we can keep track of what we are thinking?

Consider the experience of learning to read. As you sounded out each letter, you paused. In the end, you put together the sounds and sounded-out the word. What bridges the gap between the time frames of neuron activity and this behavior?

A similar issue occurs with motor neuron sequences. Consider the Chinese exercise Tai Chi. The motions of the body are slow, on a time scale much slower than the speed of neural interactions.

There must be a mechanism to hold our place in a sequence. We do not move onto the next step until an external condition completes. The neuron that holds our place is said to have Focus. A neuron with Focus is in a sub-threshold state. It does not immediately decay back to its resting state.

Consider saying the letters of the word Cat:

1. We say the letter C.
2. We wait for confirmation that C was spoken.
3. We say the letter A.
4. We wait for confirmation that the letter A was spoken.
5. We say the letter T.
6. We wait for confirmation that the letter T was spoken.

Here is a diagram of the process:

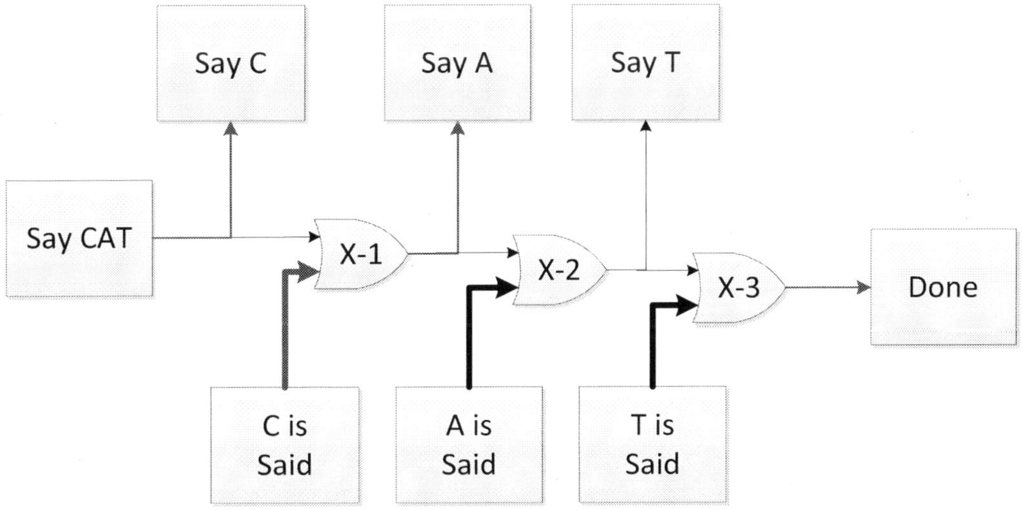

The **X** neurons represent the waiting state – where we wait for confirmation[32].

[32] With practice, we do not wait to hear the spoken letter – instead we wait for a short period of time. In that case, substitute "<letter> is said" with "wait for .1 seconds".

We need to keep the **X** neuron's voltage level from declining back to its resting state. What is required is a routinely occurring small voltage level bump. The bump would raise the voltage just a little, not enough to make it fire, but enough to keep it elevated, primed to respond.

When the confirmation signal is received, the X neuron gets a jolt. Ordinary, the jolt wouldn't cause a resting neuron to fire. But the X neuron is in an elevated state. When it receives the jolt, the effect is to push it over its threshold. It fires. The firing initiates two actions – the next letter is spoken – and the focus moves to the next neuron in the sequence.

The diagram below tracks the change in voltage of the X neuron as it transitions from

1. Receiving focus when the prior letter is complete
2. Interval decay in voltage level
3. Restorative input, maintain elevated voltage status
4. The signal that the current letter is complete;
   which raises the voltage above the threshold – the X neuron fires

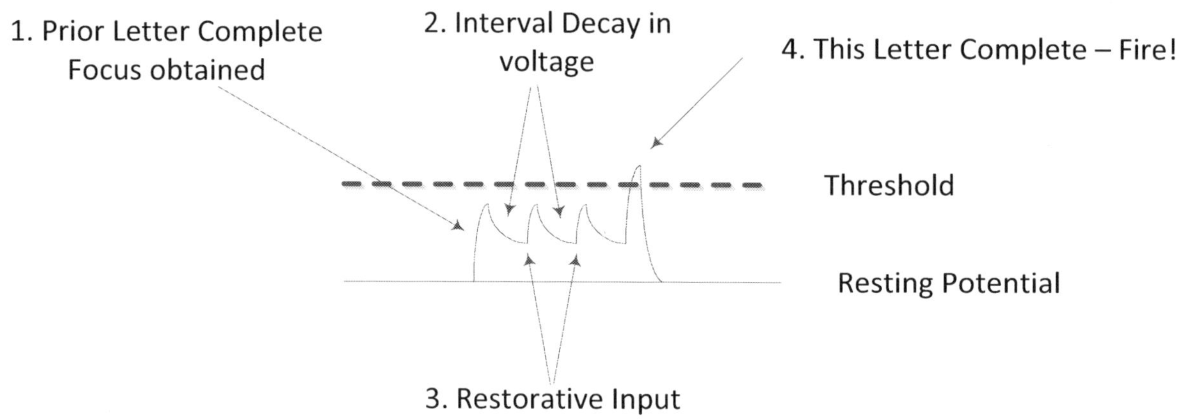

Restorative input must be regular and routine. Is there anything that we know of that could provide such a signal? The restorative input is a small bump that does not raise the voltage level of the neuron beyond its threshold.

Brain Waves

Brainwaves are synchronized electrical pulses from masses of neurons that can be measured by an Electro-Encephalogram (EEG). The frequency, or type, of the brainwave, is correlated with the state of wakefulness and attention.

Mental State	Brain Wave	Frequency in Hz or cycles per second	Wavelength in milliseconds (ms) – the time between high wave crests
Hyperactive – Active Learning	Gamma	30Hz up to 100Hz	10 ms - 33 ms
Highly Alert and Focused	Beta	14Hz up to 30Hz	33 ms - 71 ms
Relaxed but alert	Alpha	8Hz up to 14Hz	71 ms - 125 ms
Drowsy – Meditative State	Theta	4Hz up to 8Hz	125 ms - 250 ms
Deep Sleep –	Delta	.5Hz up to 4Hz	250 ms - 2 seconds

One theory is Brainwaves synchronize the neural activity across the brain. That suggests that neurons are in an elevated state, waiting for a brain wave to elevate its voltage level and cause it to fire. Instead, I suggest that brainwaves are the mechanism that prevents neurons from decaying to their resting level.

In this model, Brain waves provide a small bump to neurons with focus. Faster brain waves come along quickly enough to maintain focus. Slower brainwaves do not. The diagram below indicates that brainwaves bathe the 'X' neurons with a low level of neurotransmitter.[33]

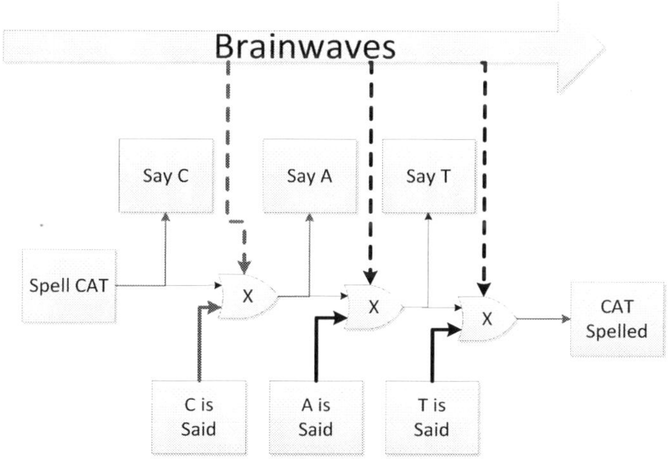

One goal of meditation is to notice when you are thinking about something and then toss it away. Meditating in alpha or theta state would require you to lose focus intentionally. There are Yogis whose meditation state is characterized by brain waves that are so slow as to prevent focus from occurring. They are in a delta wave state. The correspondence between higher frequency brain waves and conscious thought – holding on to your thought – is no accident.

[33] Note that the frequencies of brainwaves are consistent with the time constant of many pyramidal neurons, the time constant being the time it takes for a neuron to reach 63% of its final value.

Motor Sequences

The motor sequence required to spell the word cat discussed previously deserves a closer look. The firing of the Start neuron initiates the process.

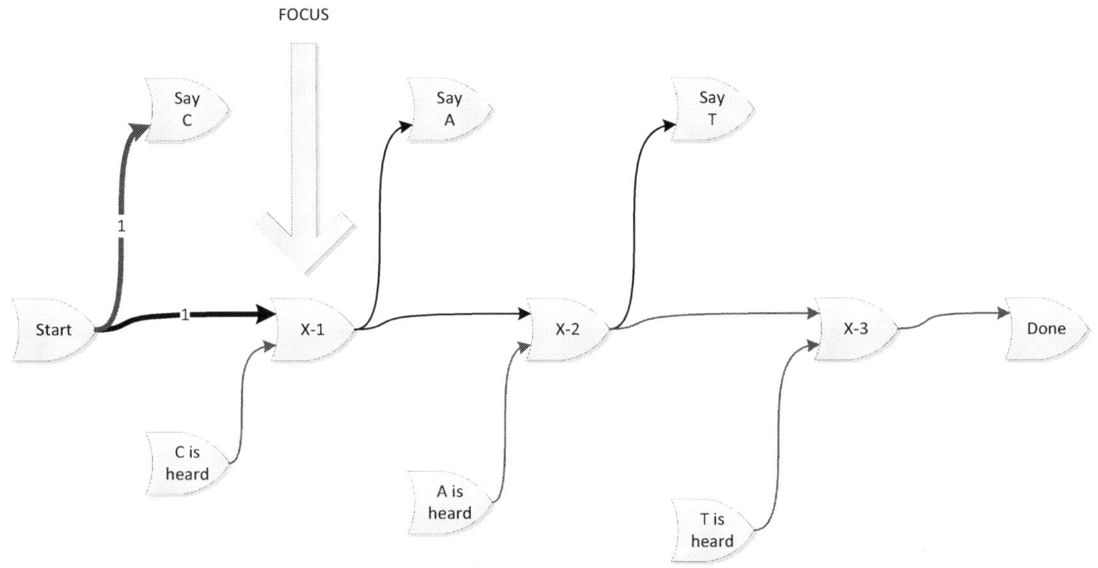

Then, the C-is-heard neuron fires, resulting in A being spoken and focus moving to X-2.

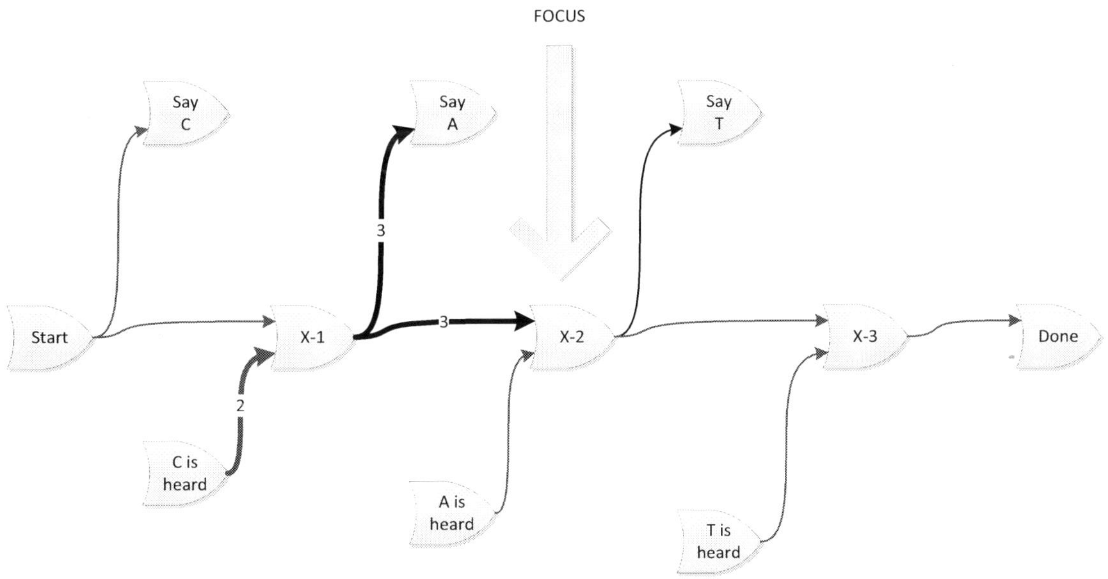

When the A-is-heard neuron fires, the letter T is spoken, and focus moves to X-3.

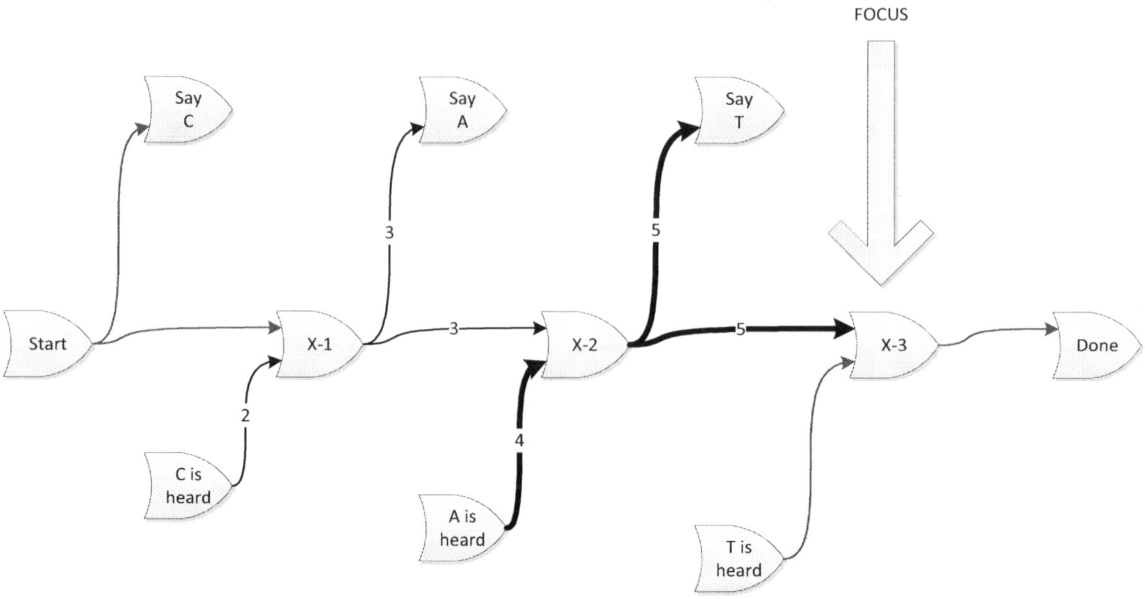

When the T-is-heard neuron fires, the process is complete.

Motor sequence creation is complex, and not that easy to follow. The next page details the steps that create a motor sequence. It is not necessary to fully understand all of the steps to go on.

Creating Motor Sequences

The mechanism for creating these neural connections is related to the predictive event mechanism described earlier in Classical Conditioning.

The process:

1. The diagram below labels the neurons as if they were Bell, Food, and Drool neurons, where one set of these three corresponds to a single letter.
2. When the Bell neuron fires, it does NOT send the Drool neuron over its threshold. Instead, it raises the voltage level of the Drool neuron enough to give it focus.
3. When the Food neuron fires, the Drool neuron is pushed over threshold, fires, and loses focus.
4. The process repeats for each letter in the word.
5. The Drool neuron of one letter's production is the Bell neuron of the next cycle.

The diagram below shows this process for two consecutive letters.

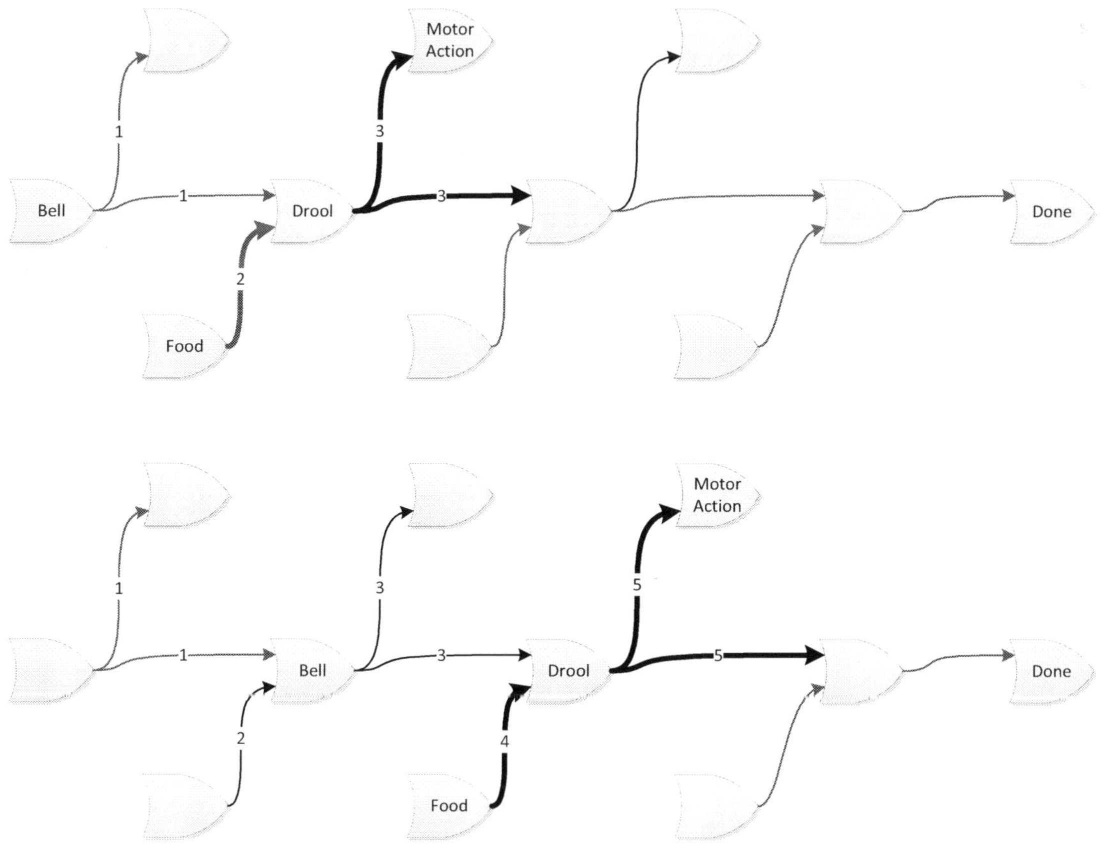

Playing a musical piece

You have heard musicians playing entire songs without referring to their printed music. They have played the piece over and over and now have the entire sequence of notes memorized. The diagram below represents the sequence of notes that they have memorized.

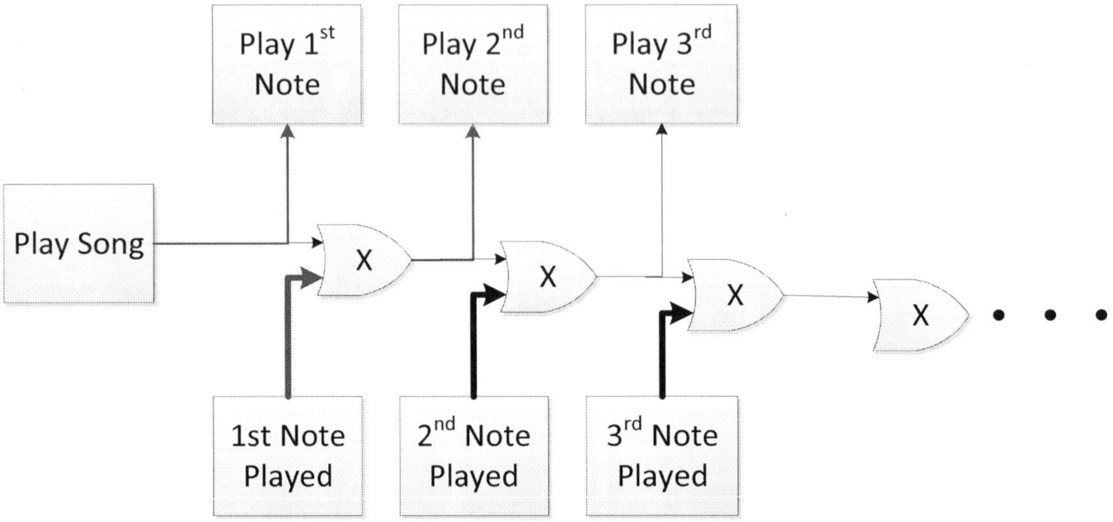

While memories of this kind are sometimes called 'muscle memory,' the memory is retained in the neocortex in a premotor area, not in the muscles.

Next, we descend to the details of playing each note.

Playing individual notes

A specific song has these four notes in sequence:

1. Play 1st note - ¼ Note Middle F
2. Play 2nd note - ½ Note Middle C
3. Play 3rd note -- ¼ Note Middle F
4. Play 4th note - ½ Note Middle C

NOTE

Each note waits for the prior one to complete.

The shaded boxes contain the muscle movements for each note. The unshaded items are the sequence of notes, not how to play them.

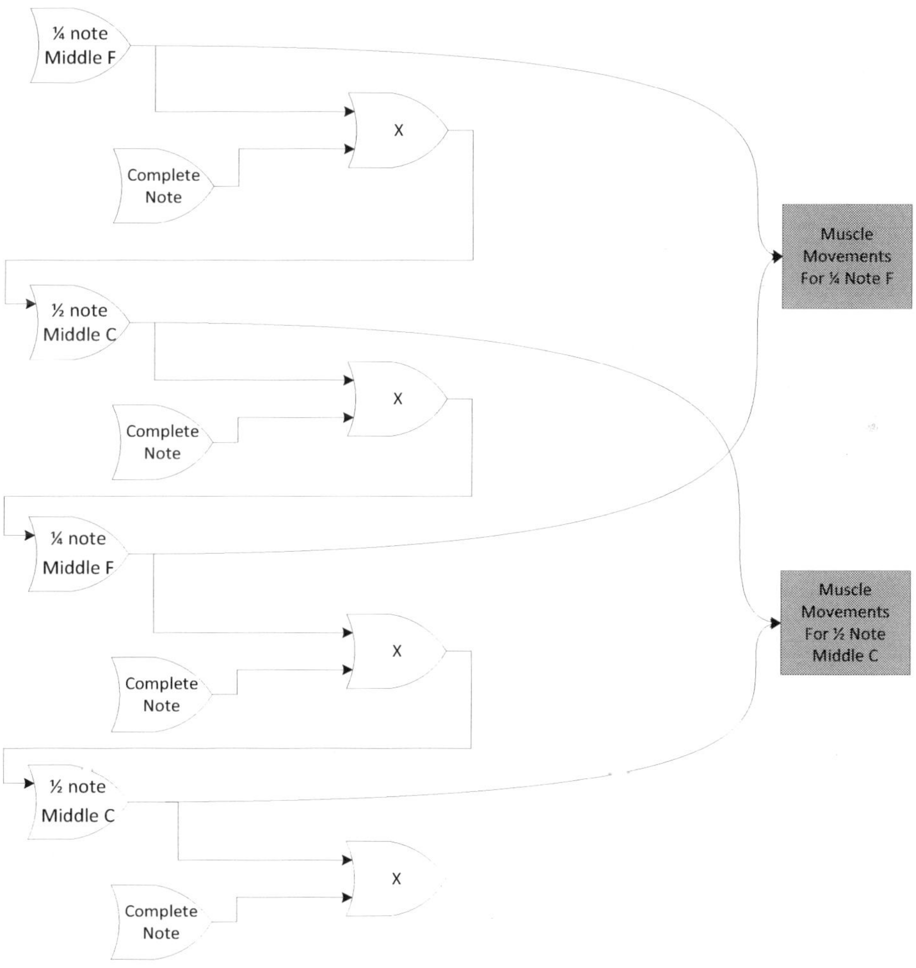

The sequence on the left is particular to the musical piece that is in memory. The shaded boxes on the right are not. They represent the movements to achieve that note for a specific duration. They exist in one place.

In the preceding diagram, there are two shaded boxes. One is labeled muscle movements for ¼ note F, and the other is labeled muscle movements for ½ note Middle C. Those boxes represent a sequence of motor neurons firing to move actual muscles.

The diagram below expands one of those shaded boxes into the movement of three muscles.[34]

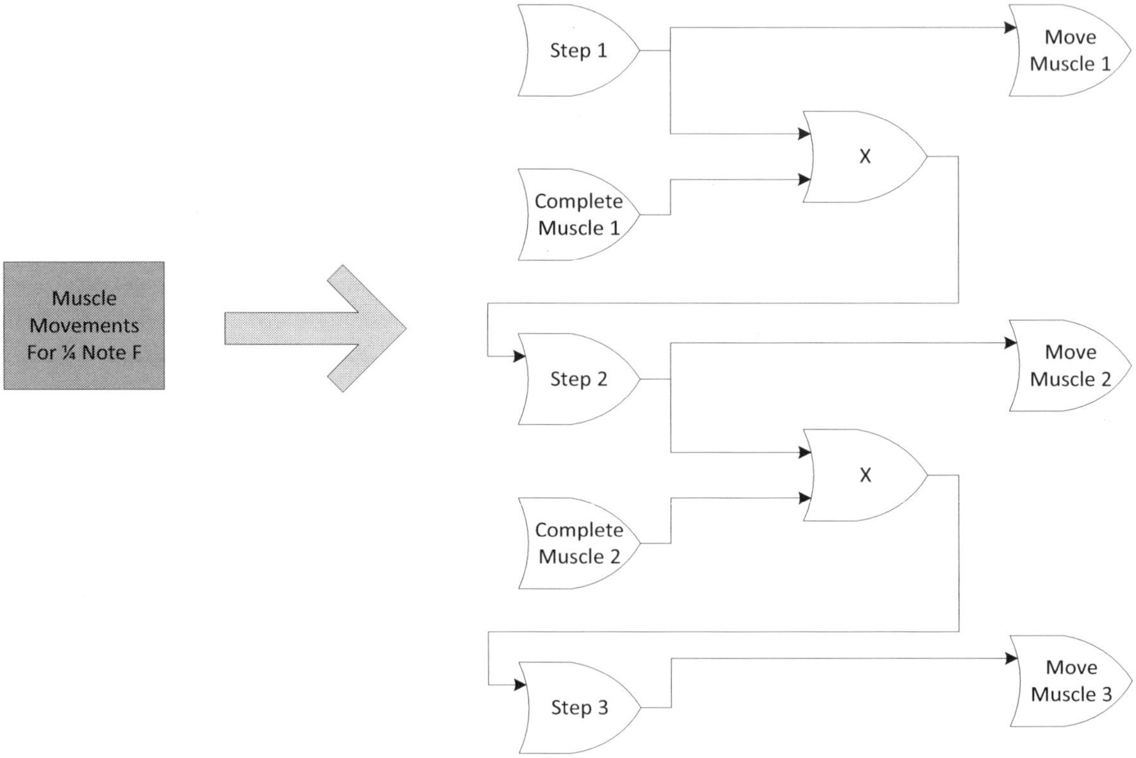

While this example is about playing music, other motor sequences are involved in activities such as writing, typing, or throwing a baseball.

The 'Move Muscle x' neurons are in the motor cortex. The other neurons are in the premotor cortex.

Order of Learning/Creating Synapses

While it is obvious that playing a song consists of playing notes, and that playing notes consists of muscle movements, it might not be obvious that we learn the process in reverse. First, one learns to move muscles to play notes. Then, you build upon that by learning sequences of notes, that individually reference the movements to play that note.

[34] The observant reader will notice that the last step does not wait for completion. The waiting for completion is found in the outer process, the motor sequence that invoked these steps.

Overall Connectivity

Before discussing the goal mechanism, it is important to consider the overall structure of the neocortex. The neurons of the association cortex are continuously firing as one thing leads to another. The diagram below illustrates this by tracing the firing of a series of numbers, numbered 1 through 10.

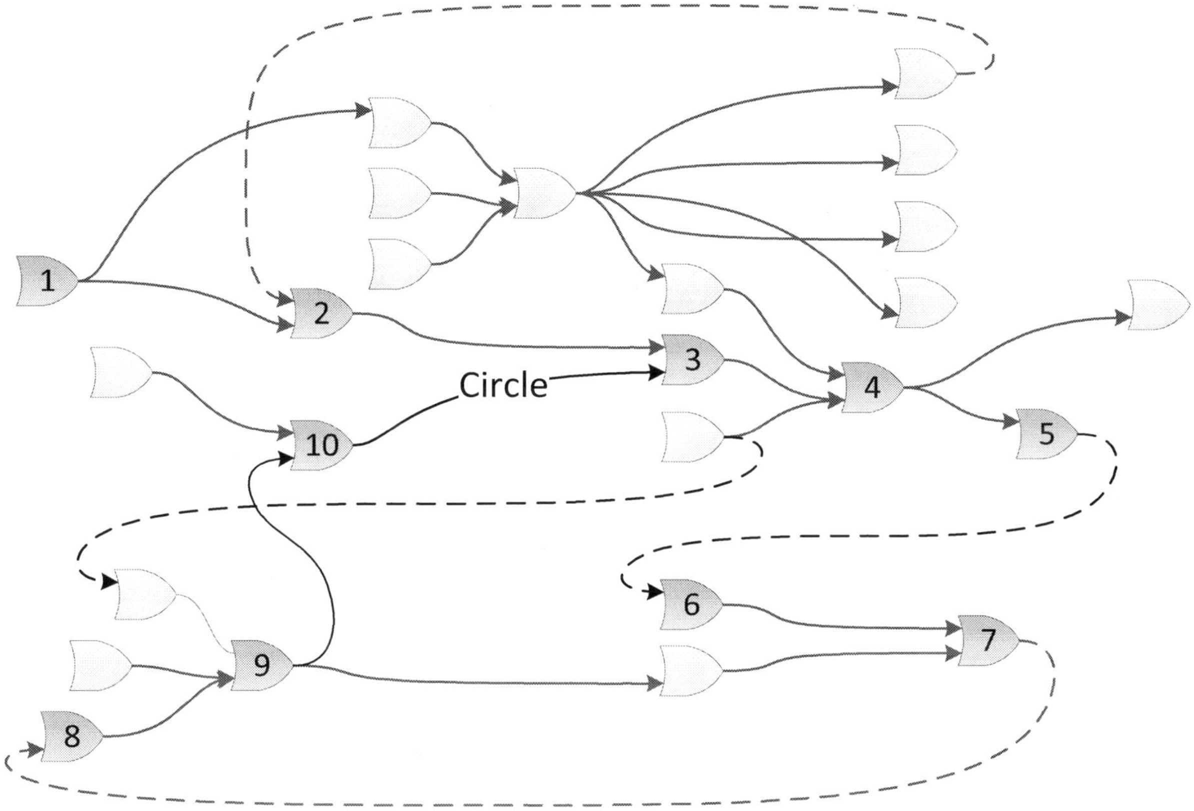

While sensory input drives firing, even when we are not attending to sensory input, if we remain conscious and are considering some goal, the firing continues. Concept neurons that are words contribute to the inner monologue as they fire.

In general, neurons receive inputs, sum them, and potentially fire over milliseconds. In motor sequences, we learned that a neuron with focus does not fire until an external event occurs. There are other sequences that fire when an internal condition is satisfied.

Other Sequences

Have you ever told someone how to accomplish a task, like changing spark plugs? In the procedure, you progress from removing the spark plug wire, removing the old spark plug, inserting a new spark plug into the engine block, and then attaching the correct wire to the spark plug. The order of steps is key to success. A procedural sequence is a sequence of concept neurons in the order of the procedure.

Another type of sequence is when you relate to a friend, the events of a movie. You start at the beginning of the movie, describing main events, and end at the conclusion. You need to relate the details in the same order that the movie unfolded, for your friend to follow your description.

The steps of these sequences do not fire into motor neurons. However, some neuroscientists have speculated that these non-motor sequences hijack the abilities of the premotor cortex.

Yet, there is an important difference between motor and non-motor sequences. It is unusual to forget a motor sequence, like, how to ride a bike. Forgetting non-motor sequences is not unusual.

Goals

Real-World Example

Before diving into the neurobiological details of this subject, a real-world example will prepare you for this otherwise abstract notion.

You are constructing a box where two pieces of wood need to be attached. The first goal is, how do you attach the two pieces of wood? The requirements are that two pieces of wood are firmly attached and do not move after they are attached, diagrammatically:

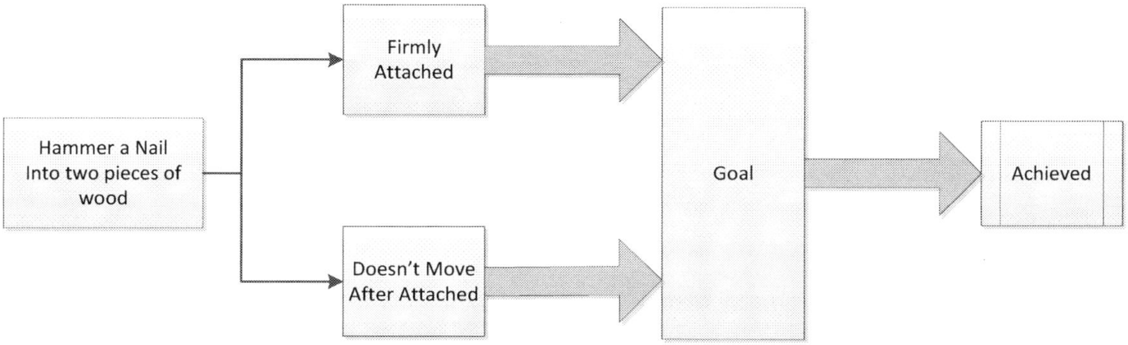

Hammering a nail into the two pieces of wood meets the requirements. The next goal is to find the hammer and nail you will be using, diagrammatically:

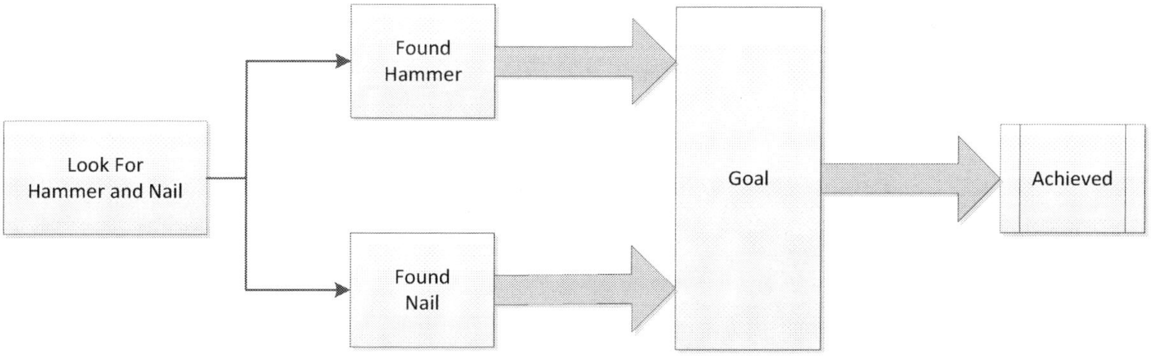

With the hammer, nail, and two pieces of wood, you hammer the nail into the two pieces of wood. Unfortunately, the wood splits. The wood is very hard, and you are stumped. How can you prepare the wood so that it does not split? You google the answer – you need to pre-drill a hole. You incorporate that into your concept neurons and connections, and now, you know how to attach the two pieces of wood without them splitting, diagrammatically:

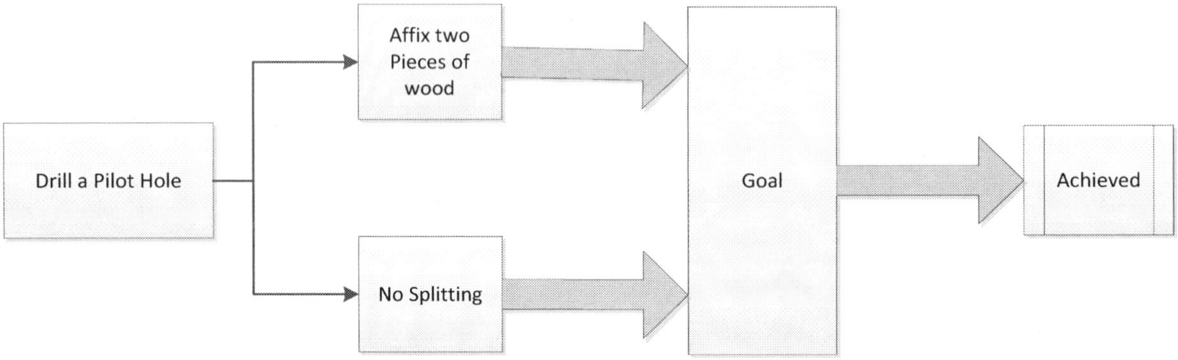

But how do you drill a pilot hole? With a drill and a drill bit. First, you find a drill and an appropriately sized drill bit, diagrammatically:

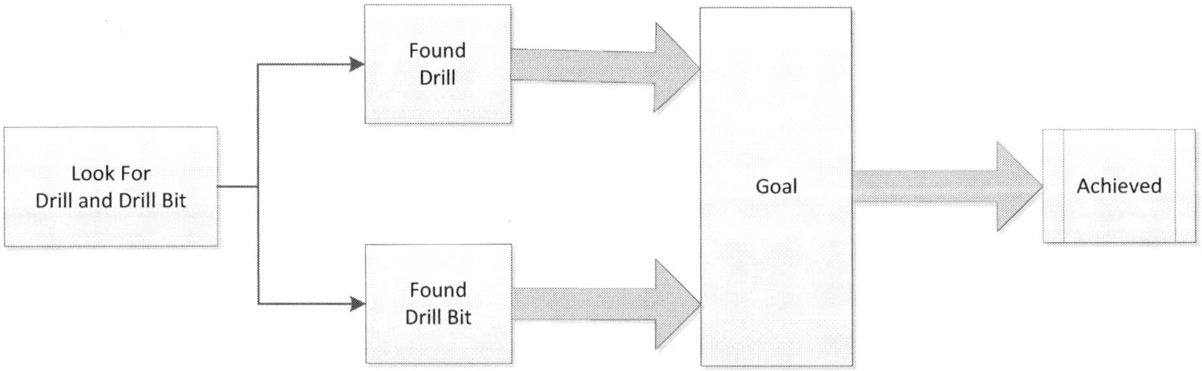

Then you drill the pilot hole.

In working your way through this matter, you have constructed a plan, a sequence of actions that are required to meet the ultimate goal, connecting the two pieces of wood. Achieving the initial goal is accomplished by working the plan, in the reverse order that the steps were determined, i.e.

1. Find the drill and drill bit,
2. Drill a pilot hole,
3. Find a hammer and nail,
4. Nail the two pieces together.

Now, on to the neurobiological mechanisms that are analogous to the above.

Goal Neurons

Sequences wait for external events to complete. Goal neurons are neurons with focus waiting for an internal event (or events) to complete. The events the goal neuron is waiting for are the requirements of the goal. A simple example of a requirement is attempting to remember something you used to know.

Ordinarily, you remember many things quickly. You readily recall the number of legs on a cat because there is a direct link between the CAT concept and the property = 4 legs.

But sometimes, you can't remember something that you are sure you used to know. For example:

You are talking with your friends about favorite TV shows. The subject turns to the TV show "Friends." One of your friends says that Chandler was her favorite character. Another friend says that Joey was his favorite character. Your favorite is Rachel. You are trying to think of the actress's real name but can't remember it.

You are sure you knew her name at one time, but you can't think of it now. A few minutes pass, the discussion goes on to other subjects. Suddenly, the answer pops into your head. You say out loud, Jennifer Aniston.

The diagram below represents the status of the related neurons before you began trying to think of the answer. Note that the dashed connection represents a connection that used to exist.

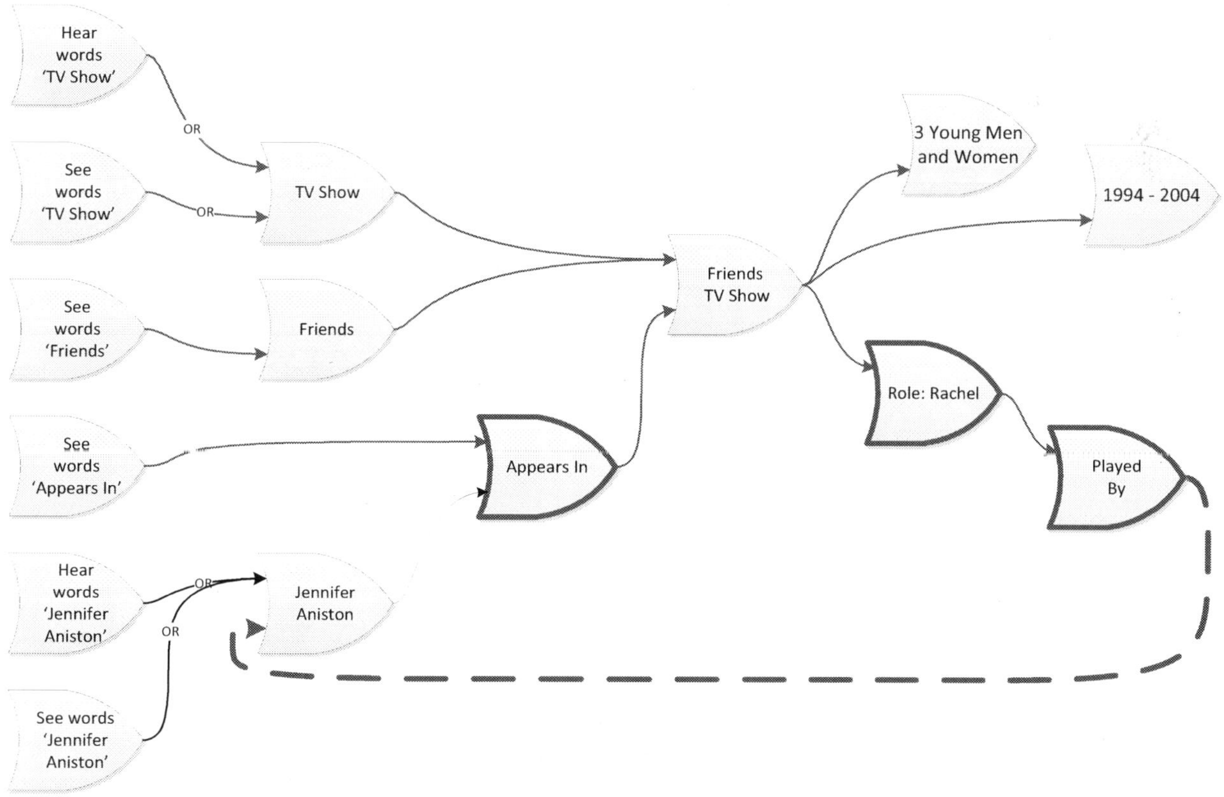

The dashed connection reflects the fact that you used to know who played the role of Rachel. But after little or no reinforcement of the synaptic connection, that pathway faded.

Below a goal-neuron represents realizing the forgotten detail. When the requirement of the goal is satisfied, the goal neuron will fire, as the answer becomes clear.

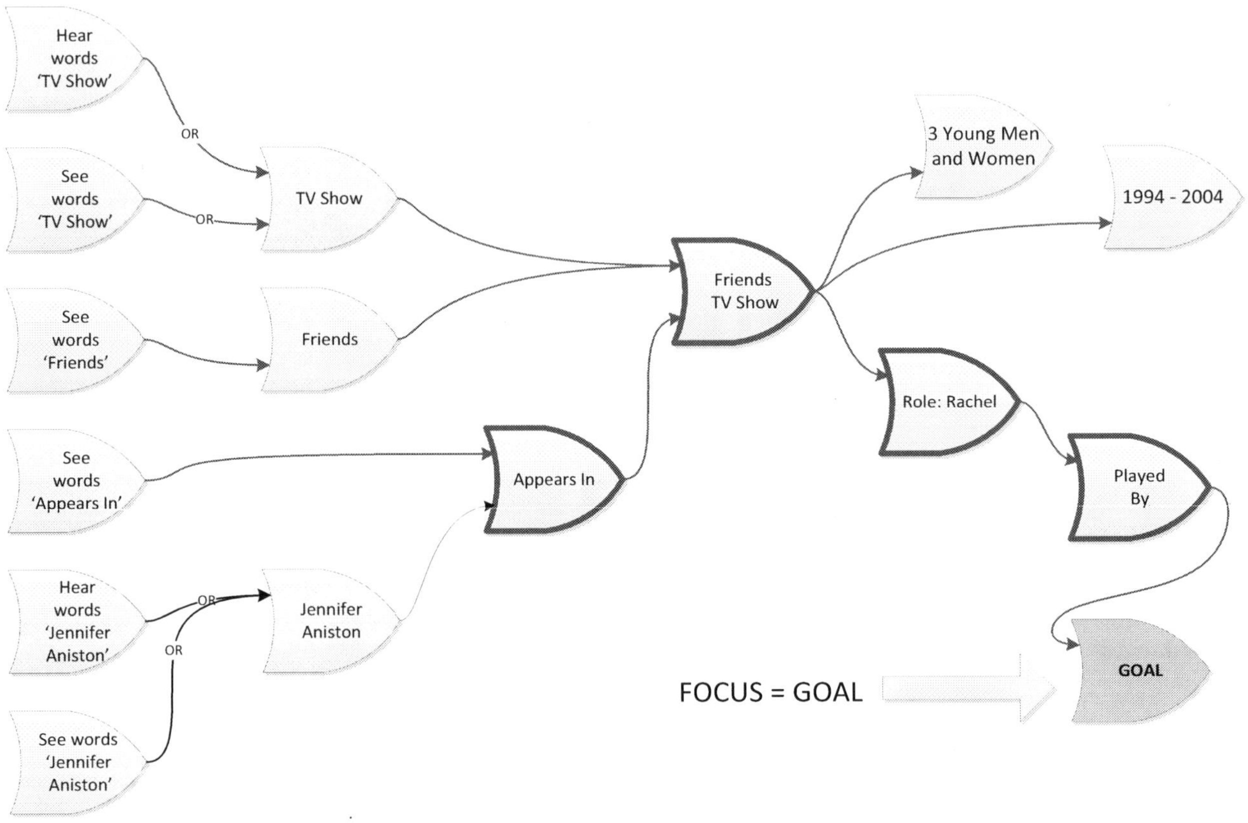

After the conversation moved onto other matters, you no longer consciously thought about the matter. But in your web of connections, where one thing led to another when the Jennifer Aniston neuron happened to fire, its property neurons also fired, i.e.

Jennifer Aniston => Appears In => Friends TV Show => Role: Rachel => Played By => GOAL

And with that, the answer popped into your head. The firing of the goal neuron indicates you found an indirect connection.

Repeating the words "Rachel was played by Jennifer Aniston," over and over, will form and strengthen the re-established synapse connection shown below.

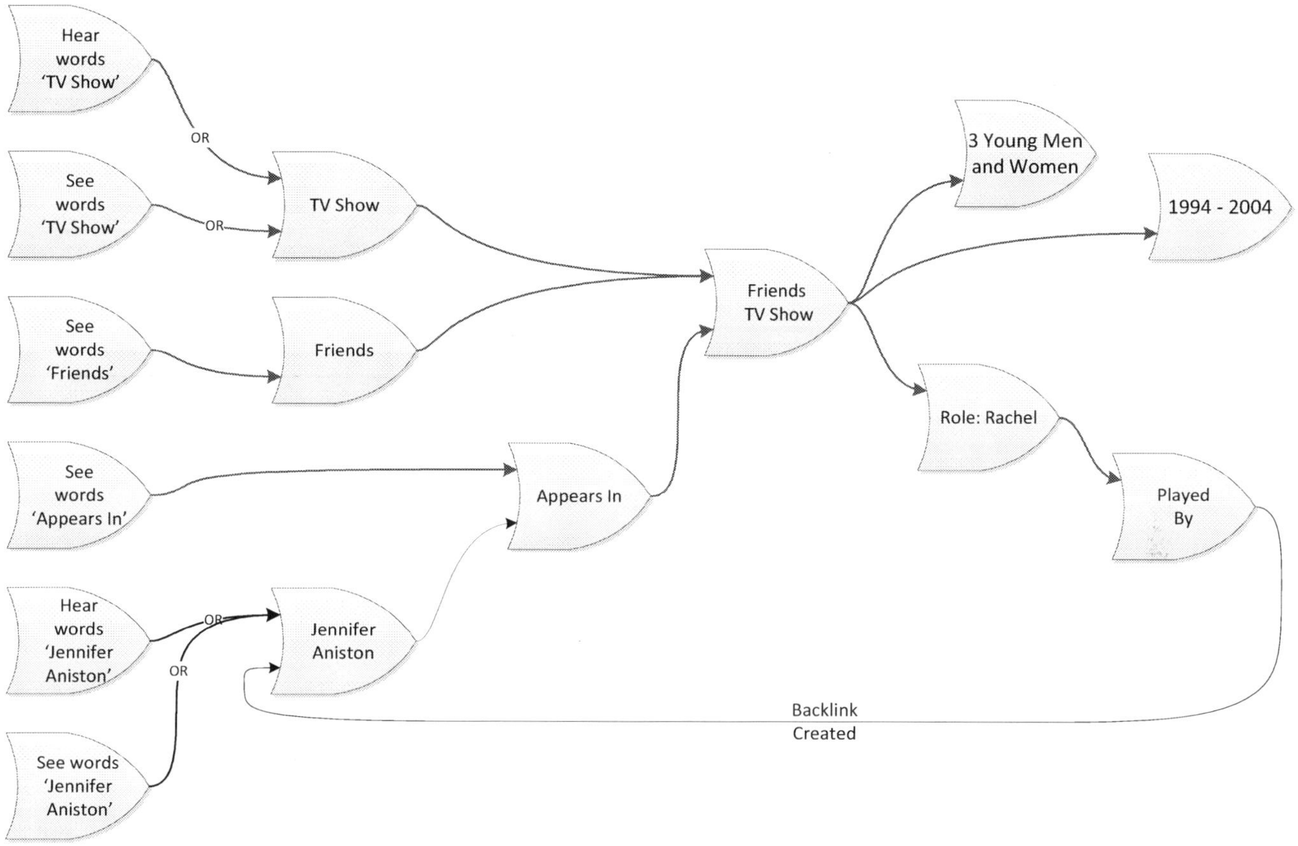

Once the connection exists, you will be able to answer the question "Who played Rachel on Friends?" quickly, without wandering in your web of concept neuron connections.

In this example, we are dealing with a fundamental underlying aspect of neural connections – they are essentially one way. The feedback connection above overcomes this limitation. The mechanism that creates that connection is explored immediately after the next topic, explicit versus implicit knowledge.[35]

---

[35] At this point I am unable to provide a definitive explanation for how the backlink connection from the goal neuron identifies the proper neuron to connect to.

Explicit versus Recognition Knowledge

To better understand the implications of the link, go back to the point where the new link didn't exist. Can you answer the multiple-choice question?

"Who played Rachel on the TV Show Friends?"

(A) Courtney Cox
(B) Lisa Kudrow
(C) Jennifer Aniston

Yes. You can locate each of the three concept neurons and follow their properties. When you got to answer (C), when you try the Jennifer Aniston choice, her property links fire in turn, i.e.

Jennifer Aniston => Appears In => "Friends" TV Show => Role: Rachel

Recognition knowledge is the ability to recognize the correct answer. Explicit knowledge is required to answer the question:

"Who played Rachel on the TV Show Friends?"

Only with the newly recreated backlink connection above can you quickly answer this second question. This ability, to answer the second question, requires that you know the subject "backward and forwards."

Two Requirements for a Goal

The above example had only one requirement neuron feeding into the goal neuron. The example below includes two requirement neurons feeding into the goal neuron. The goal neuron fires only after both requirement neurons fire. You are trying to locate a red hammer.

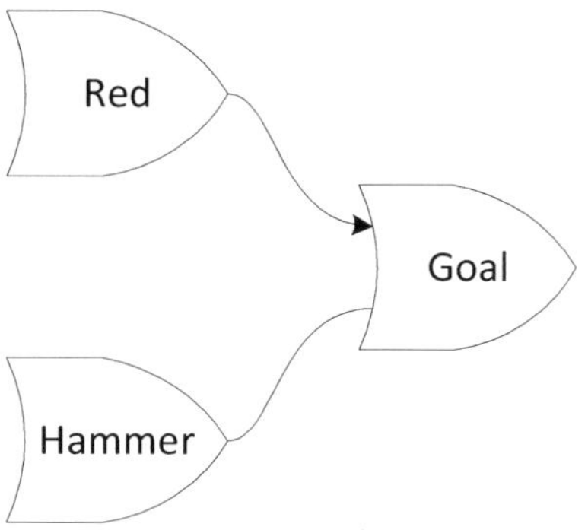

As you think about, you recall a red hammer, i.e.

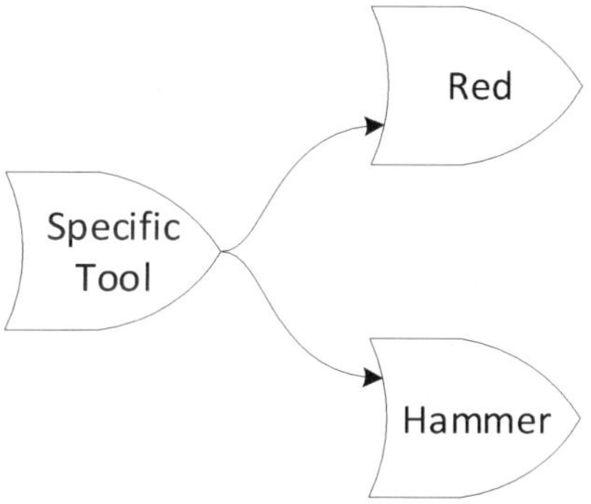

As the tool neuron fires, the cascade of neurons firing satisfies the goal. The property of the Tool neuron tells us that the specific tool, the Red Hammer, is in the basement closet.

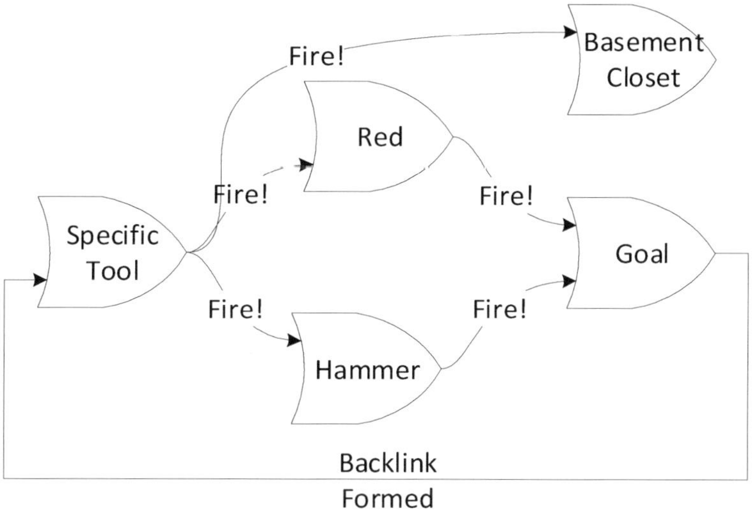

Iterating Towards the Goal

Imagine you are a groom on the way to your wedding. Suddenly, the traffic in front of you slows down and becomes stop and go. You can't afford the delay; you have to get to the church on time[36]. You didn't bring your GPS. What now?

Fortunately, you used to be a taxi driver and know the area well. The diagram below represents your knowledge of roads in the area as neural connections. The big arrow shows where you are now. Note that all the connections between neurons are one-way connections. For example, the road connection location A and B are one-way. The road connecting B with E goes in both directions as it has both feedforward and feedback connections.

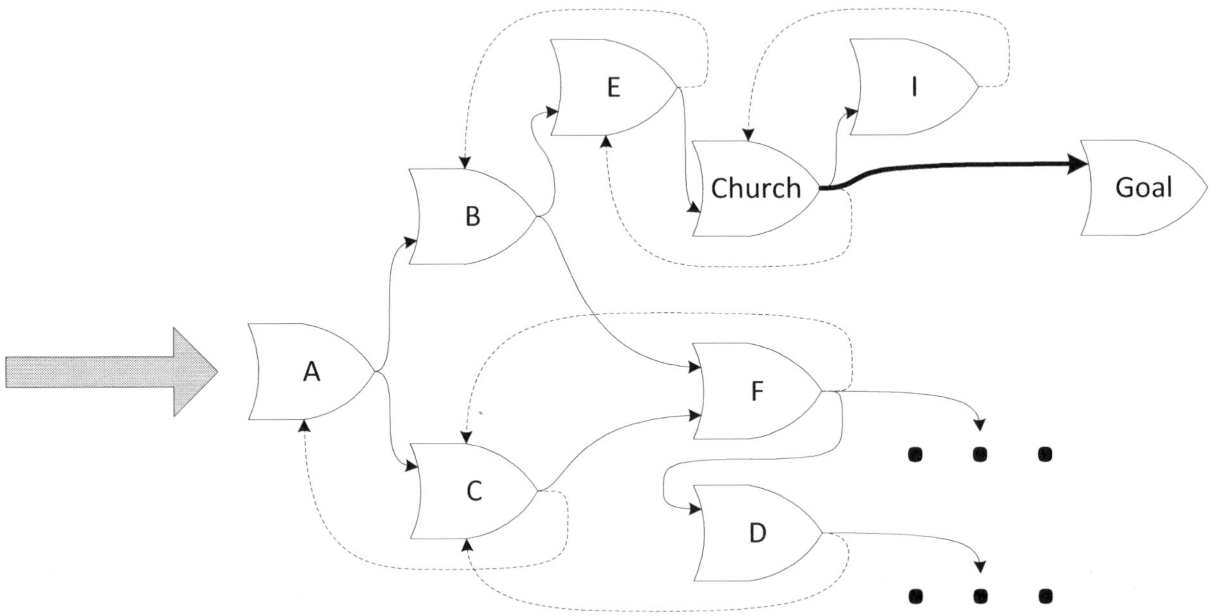

Following your neural connections reflect roadways that run between locations. There are both solid lines (feedforward) and dashed lines (feedback).

In that you are now at location A, as it fires, you traverse the neural connections A->C->F->D->C and also A->B->E->Church-> Goal.

As per above, when satisfied, the goal creates a backward link to the solution neuron, which is where you travel to next - location B.

[36] Apologies to David Bowie (Modern Love) and to Lerner and Lowe (My Fair Lady).

When arriving at B, as per the diagram below, you once again think of your goal, and the connected neurons light up in sequence, i.e., B->F->D->... and B->E->Church->Goal.

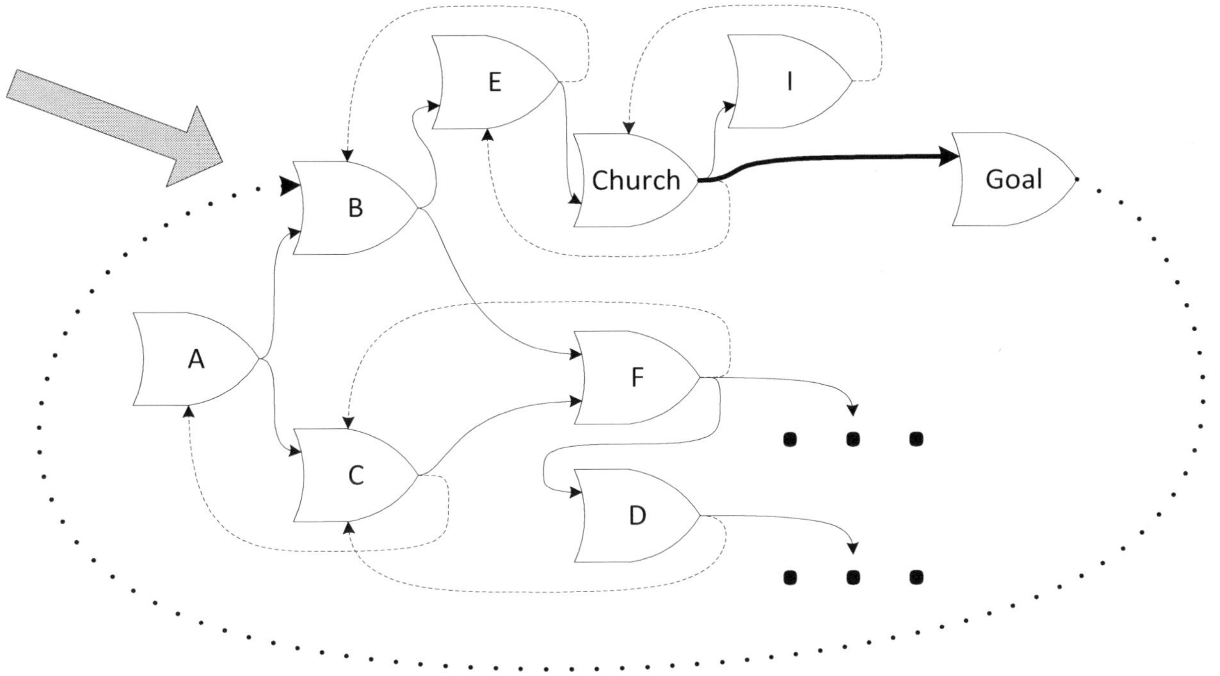

Again, the goal neuron creates a connection back to the solution neuron, which is E, as per below. You then travel to location E.

The situation when you arrive at location E is as pictured below.

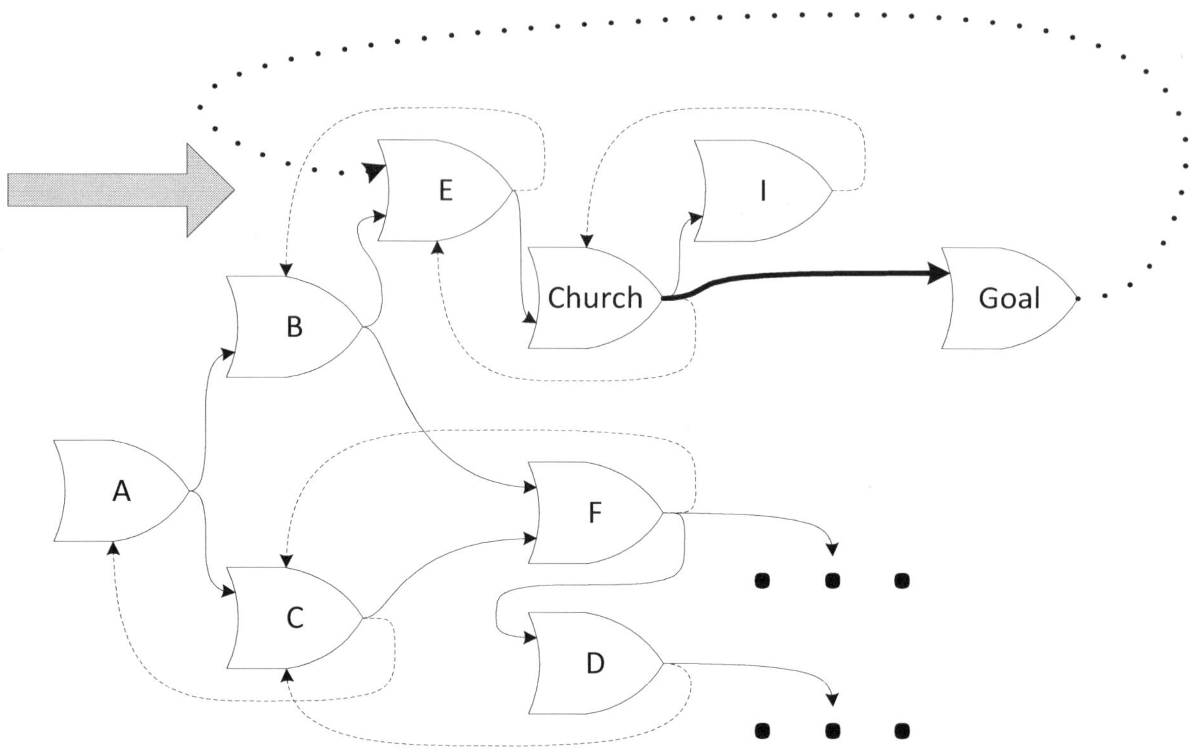

From location E, you can see the Church in the distance; you drive directly to it.

Iterative refinement solutions are key in how we interact with the physical world. As we walk, we adapt to the surface we are walking on when on a boat we adapt to the rocking of the boat, and as we catch a ball, we adjust our actions as the wind blows the ball.

In attempts to create robots that perform like humans, computer programming performs complex calculations to determine precisely how to move actuators or spin motors. That is not how we move. We continuously access our state, adjusting our movements. When we are moving too fast, we slow down. If we need to jump higher, we do. No numerical calculations are involved.

Iterative refinement allows us to continuously adjust our actions in response to a continuously changing world.

Sub-goals and External Knowledge

In the above example, the real-time aspect meant you could redefine the goal successively as you approached it.

While planning for the future, you need to break down the goal into steps – sub-goals – that, when performed in sequence, reach the final goal. For example:

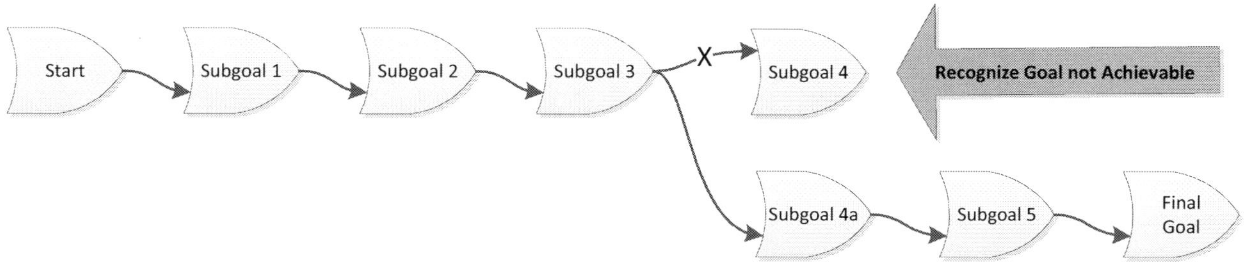

As illustrated above, after formulating a plan to achieve Sub-goal 4, it was recognized that there was no pathway from there to the 'Final Goal.' So, the process had to back up one step. The next trial started as Subgoal 4a[37].

As plans and details become more complex, we resort to external records. Similarly, we can access external sources of information to identify solutions. In both of these cases, language is the medium.

Obtaining external knowledge creates new synapses that, when traversed, satisfy the subgoal.

---

[37] This illustrates the common-sense expression: "Plan your work, Work your plan."

Goal Process Flowchart

The process for creating Goals and Sub-goals is flow-charted below. The box labeled 'Directed Search' involves gating neurons that direct search to closely related areas. When you run out of patience, you can abandon the directed search or proceed with a background search. Background searches proceed without direction from gating neurons.

Sub-goal Sequence Example – Creating a Wooden Box

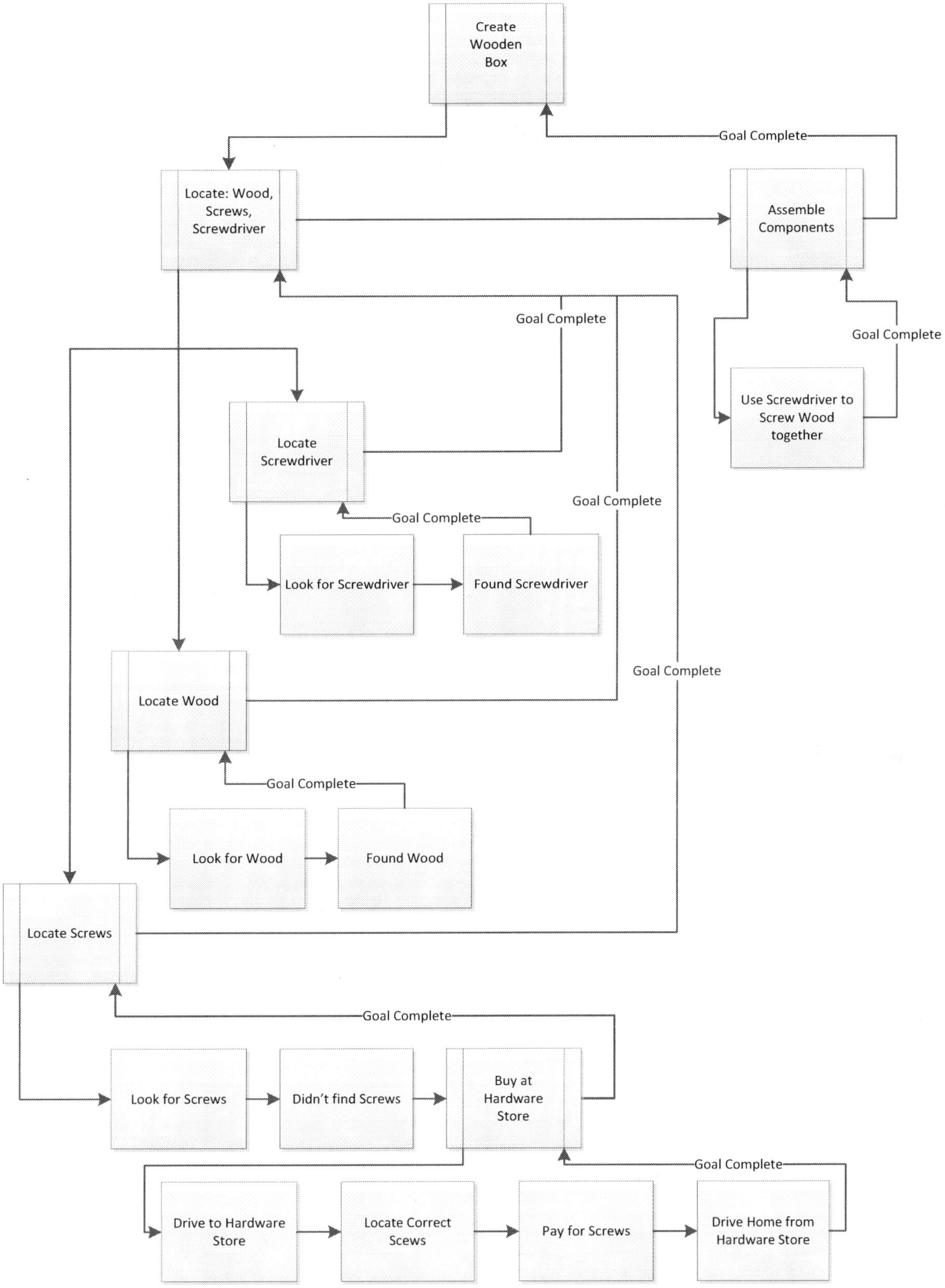

Automatic Behavior

Simple Automatic Behavior Example

The example below is from a new driver learning how to drive. The vehicle is moving quickly, and the road is turning. The new driver doesn't know what to do yet. Fortunately, the instructor tells the student to turn the wheel to the right gradually.

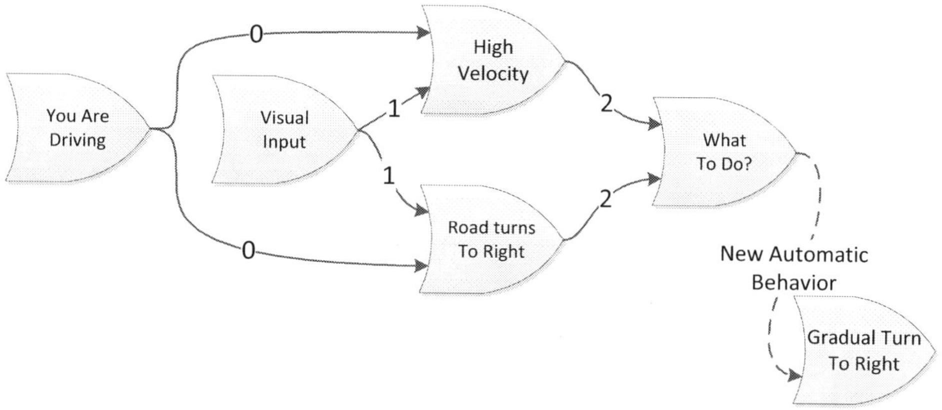

The student turns the wheel gradually, the goal is satisfied, and new behavior – an automatic behavior – is established.

Because the result was a good one, the brain secreted dopamine. Dopamine helps establish new synaptic connections, a connection from the 'What to Do' goal to the action that satisfies that condition/goal.

Establishing Automatic Behavior

Connections to motor neurons are required to satisfy goals that require physical movement. The initial attempts to satisfy such a goal may be inadequate. But, with repeated experimentation, successful and unsuccessful methods are discovered and incorporated into neural circuitry. Exposure to diverse initial conditions expands the scope of learned responses. With continued practice, iterative refinement improves the response, shaping future behavior.

Consider the major league baseball player who is playing in right field. This professional player has the skills to quickly predict where the ball is going and move in that direction to catch it.

That player was not born with that ability. Instead, like the rest of us, the player first learned to catch a ball when a partner gently tossed a ball from a short distance. The initial attempts to catch the ball fell short. Either the player moved in the wrong direction or was too late to catch the ball. In the beginning, the player made guesses about how to move their arms to catch the ball.

After enough attempts, the player discovered how to be successful. Then the distance between the thrower and the catcher is increased. Again, after enough trials, the player discovers how to catch a ball thrown from a further distance.

The player comes to understand that you must incorporate how the ball is released, how quickly it rises or falls, and whether it curves to the right or left. Each variation leads to new automatic behaviors.

The major league baseball player in right field went way beyond catching a ball in the backyard. But the principle of how the player learned to catch balls in the outfield is the same. With practice, the major league baseball player constructed a set of automatic behaviors that, when combined, enable him to catch a ball way out in the right-field. Without any thought, automatic behavior is selected based on current observation of the initial trajectory of the ball, the ball's rate of climb, the wind, which leads to running to where the ball will land. As the ball continues to travel, so does the player, adjusting with each moment in a series of automatic behaviors.

Soldiers train and train – over and over again – to make automatic their reactions in life and death circumstances. They can't afford to figure out what to do when confronted with a new circumstance; they need instant automatic behaviors to react in these dangerous situations. During the training periods, they are in a heightened state, forming new synapses that wire-in their newly learned automatic behavior. In each new circumstance, there is a goal that reflects the current circumstance and the desire to remain alive. The training circumstances light up the goal and create/reinforce the automatic behavior.

Automatic behavior has several advantages. First, the response is quick – much faster than the thought process requires. Second, the number of neurons involved keeps metabolic costs down. And, you can direct your attention to another matter entirely. You can talk, listen to the radio, or daydream without negative consequences.

Automatic behavior conditions are specific to the current environment. For example, only if you are driving are the automatic behaviors relevant to driving operable.

What Brain-machine interfaces have told us

The April 2019 Scientific American article 'The Intention Machine,' by Richard Anderson describes his lab's work on brain-machine interfaces (BMI). Placed on a subject is a neural activity monitor. The neural activity monitor records the activity of individual neurons. In particular, an area of the brain that initiates limb movement, a region of the Posterior Parietal Cortex (PPC), is monitored. Richard Anderson writes:

> People often ask how long it takes to learn to use a BMI. The technology worked right out of the box. It was intuitive and easy to use the brain's intention signals to control the robotic arm. By imagining different actions, the subject, Sorto, could watch recordings of individual neurons from his cortex and turn them on and off at will.

The neurons monitored are at the beginning of the cognitive command chain that initiates limb movements. The observed neural activity is directly associated with a specific limb movement goal. The neurons that fire into motor neurons are downstream of these 'intent' neurons.

These 'intent' neurons are a form of concept neurons. Before these concept neuron's firing could initiate an action, the individual had a goal, for example, to move a limb. When that goal neuron fires, the goal is achieved. After that, the intent neuron's firing results in the limb moving.

Mirroring – Automatic Behavior from Goals

In a fully developed human being, the same neuron, a mirror neuron, fires whether the human is performing an action, or observing another performing the same action.

In that, no one is born with mirror neurons able to perform these functions - how are mirror neuron connections created?

Newborns fire motor neurons in what appears to be a random pattern. They are not able to consciously control their movements, as there is no connection from an intent concept neuron to the motor neuron or neurons that they are firing. Establishing that connection is key to forming 'mirror neurons.'

Newborns watch others move their right hand. The newborn sees this often enough that it now perceives the concept of moving a right hand. Of course, the newborn has no idea how to accomplish moving their right hand, initially.

Imagine the newborn focusing on moving their right hand – meaning it is a goal to see the right-hand move. The diagram below shows the initial state of neural connections.

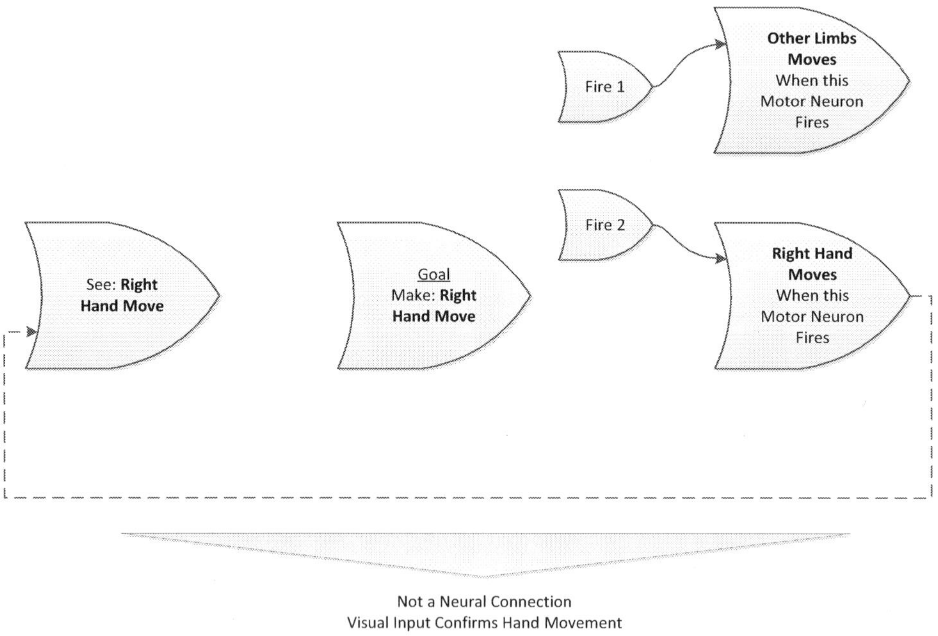

Initially, there is no connection between the Goal to make the right-hand move and the neuron that moves it. The dashed line that runs from the "Right Hand Moves" neuron to the 'See Right Hand Move" neuron reflects the fact that the newborn perceives when their right-hand moves, firing the 'See Right-Hand move" neuron when their Right-Hand moves.

As the newborn randomly fires its motor neurons, moving its arms and legs, it happens to fire the motor neurons that move its right hand. The newborn observes the hand moving, which means the 'See: Right Hand Move' neuron fires. Its firing trips the goal neuron, which has focus, into firing.

As the goal neuron fires, it makes the connection to the recently elevated 'Fire 2' neuron in the diagram below.

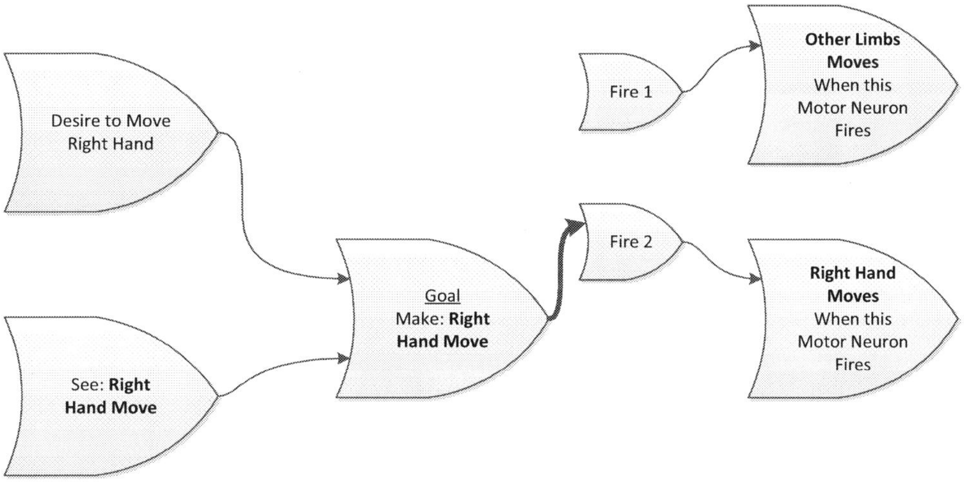

Multiple repetitions reinforce these connections.

This process occurs over and over until the infant has built up a repertoire of movements. When complete, the infant can decide how it wants to move and then move.

Deer in the Headlights

When we are daydreaming, recalling memories, observing what is around us, we are in a lower state of neural activity – one where automatic behavior is running the show. We are following well-worn pathways that direct our behavior.

What happens when a novel circumstance arises? We break out of our current automatic behavior, looking for a non-existent automatic response. Momentarily we are frozen, unable to respond. When we fully recognize that no automatic behavior is appropriate, we transition to a more active state.

In this state, we must quickly decide what to do. The circumstances could be how to: avoid a collision, respond to an unheard teacher's question, or avoid tumbling down a hill. To respond, we need to quickly establish a goal whose requirements are a path from the current circumstance to a safe outcome. The heightened state means that there is concurrent neural activity searching up and down the web of connections.

With luck, we have at least one not-terrible resolution that lights the what-to-do-next goal neuron.

Why Goals?

What motivates the creation of a Goal? What motivates any action? The four F's are a source. When a creature needs/wants to Feed, Fight, Flee or reproduce, a neocortex goal is required.

These drives arise from subcortical neurons, and there are neural connections into the neocortex to establish goals to meet these needs.

Alternatively, goals also originate in the neocortex. As social creatures, we have goals to maintain or elevate our social status. A higher social status implies an improved chance of survival.

And, importantly, there is the fifth 'F,' fun. With a large association cortex that is initially empty, Fun is a way to populate the knowledge links that help us survive. Fun can be the play that occurs between siblings. In humans, education needs to be fun.

Up Next

An important differentiation from other animals – language. The ability to inform with great specificity and persistence in time (writing).

Language

Before (Verbal) Language

When my dog walks over to the door to the yard and looks back at me, he communicates what he wants. When he sits near the dog treat jar, looking at it, then looking at me, I know what he wants. When I sit down to put on my shoes next to the dog leash, he repeatedly looks at it, and me. He is communicating that he wants to go for a walk. He does not need words.

Not all humans have verbal-language. A deaf man who never learned to sign was able to communicate his intentions and desires, not just to those of us with language, but with other individuals without verbal-language.

Human language can communicate not just what is true now, but what was true in the past, or what we want in the future. It is key to our advantages over other species. Words, of course, are the elemental components of language and explored below.

Recognizing Words

In the puzzle below, we read scrambled words with ease. There is little need for conscious thought.

How do we recognize words when reading? This puzzle shows that it isn't a simple matter of comparing what you read against all of the words you know[38]:

> **Aoccdrnig to a rscheearch at Cmabrigde Uinervtisy, it deosn't mttaer in waht oredr the ltteers in a wrod are, the olny iprmoetnt tihng is taht the frist and lsat ltteer be at the rghit pclae. The rset can be a toatl mses and you can sitll raed it wouthit porbelm. Tihs is bcuseae the human mnid deos not raed ervey lteter by istlef, but the wrod as a wlohe.**

I was suspicious of the assertion that both the first and last letters had to be in the right place. I took all the misspelled words and put them through the letter scrambling engine at https://new.wordsmith.org/anagram/. This program compares the scrambled results with a dictionary of English language words.

For example, the program produces both what and thaw when given waht. Only the first letter has to be in the right place – in all cases – but one. The exception is the word does. As I read the sentence above with the word deos in it, I catch myself uttering both does and dose quietly. That doesn't happen with any other words, so I suspect that the first letter in the right place is the absolute requirement.

The mechanism that operates to recognize words quickly compares the scrambled letters to the words we know. The puzzle above shows that rearranging the letters doesn't confuse the recognition process. The mechanism the recognizes words must not be sensitive to the order of letters in a word.

---

38 http://www.mrc-cbu.cam.ac.uk/people/matt.davis/cmabridge/

In one way, the above puzzle is like finding the answers to crossword puzzles. While in the middle of solving a crossword puzzle, entries are only partially filled in.

Often, with just a few letters, the answer to the question pops into your head. It seems as if you have an index that – given a few letters – narrows down the possible answers very quickly.

The same kind of index would help with the letter re-arrangement puzzle above.

In the speculative diagram below, there are two components. One makes sure we don't consider words that include specific letters; the other side makes sure we consider words that have all the letters in the clue.

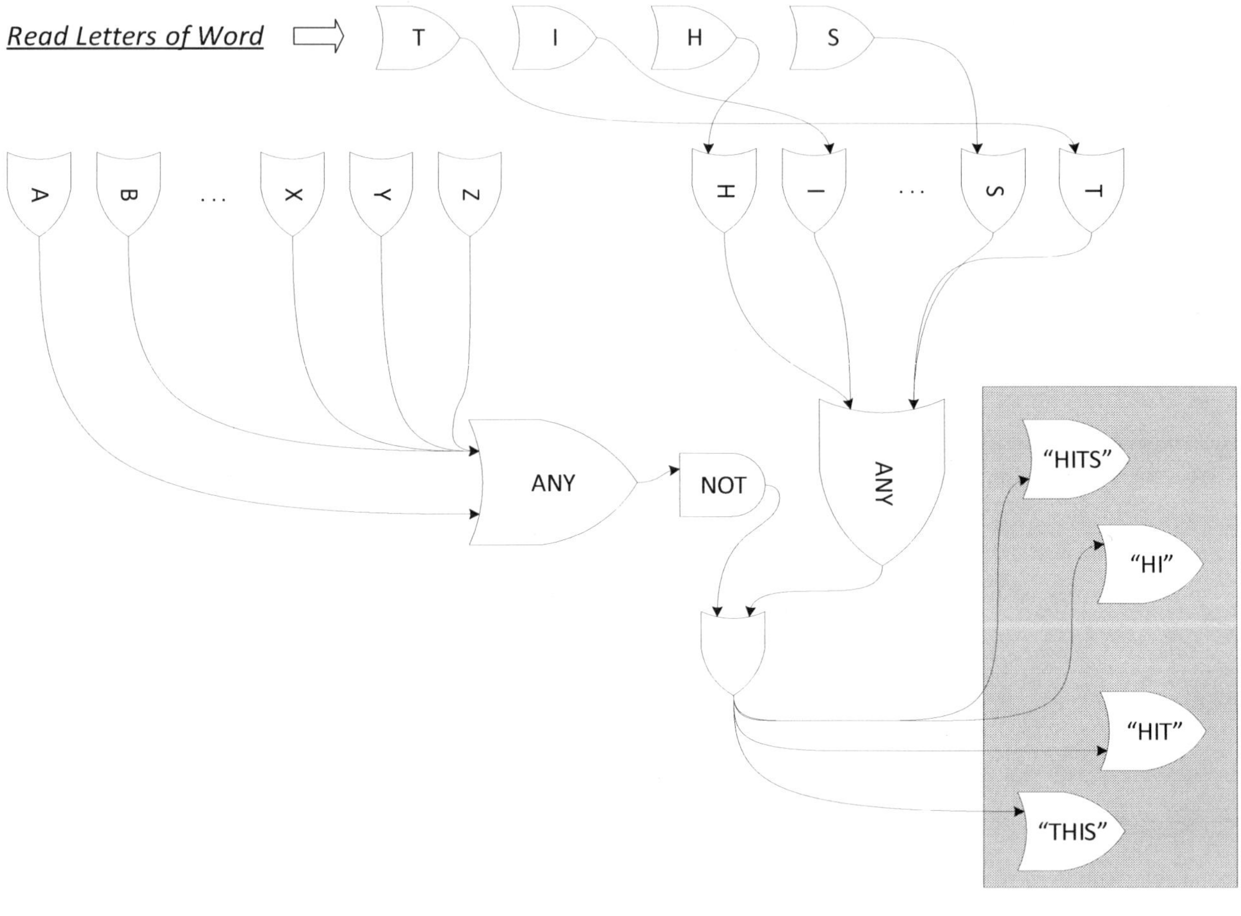

Note that there is another word made up of the letters H, T, I, and S not shown in the shaded box[39].

This kind of recognition may be present in other areas of the neocortex. It combines what is present, with what is not present, in the recognition process. Expectation, as suggested by the Prediction error mechanism, is likely involved.

Note that there is additional processing not on this diagram that selects the most appropriate word.

---

[39] This word is an acronym for 'So Happy It's Thursday."

Not Everyone is Wired the Same Way

While different people have specific differences in neural connections, two reviewers of this work indicated they read words differently than the proposed mechanism.

My friend Carl found it easier to read the scrambled word text than non-scrambled texts. Carl had a particularly hard time learning to read. In his later years, he came to suspect that he might have dyslexia.

One hypothesis is that dyslexia is a timing issue, with letters received out of order. Individuals who adopt coping mechanisms for dyslexia, like Carl, will be better equipped to deal with scrambled words. Carl has become quite the Scrabble player, perhaps in part because of his enhanced ability to identify words from scrambled letters.

Another friend, Harvey, tells me that he does not read individual words – he reads blocks of words, including common word combinations/sequences. In Elementary school, while learning to read, he was a participant in an experiment. The experiment flashed blocks of words at a time. He reads blocks of words at a time. Nonetheless, others, like his mother, a voracious reader, might read that way as well.

Only after thinking about the scrambled text quiz did I construct the neural connection diagram above. Note that additional processing that incorporates context is needed to select which of the possible words is the correct one.

The above affirms Harvey's reading process ideas. To select the correct word as you are reading it, you must have already looked ahead, gulping down the rest of the sentence - unconsciously.

Representation of the Spelling and Speaking of Words

A word that is a concept neuron has one of its properties the spelling of the word. The sequence of letters is critical. When such a concept neuron fires, the sequence of neurons that spell out the word fires as well. Unless the output is being gated, or directed, to language processing, the firing goes nowhere.

A similar requirement exists for recognizing words while reading. A sequence of letters leads to concept neuron(s)[40]. These letters do not connect to motor neurons. Instead, they are identifiers of concept neurons.

Thus, the spelling of a word appears in two places – one for identifying the word, the other for output.

In addition to the written word, speech involves both speaking and hearing words. Is there one place for each word that contains how to say and speak a word? Or, is it like the written word, where there are two distinct mechanisms?

I learned to read by sounding out words. Enhancing the ability to recognize the spoken word came along for the ride.

Or consider music. I can recognize many musical titles. But I am unable to reproduce either the words or the tune. Musicians can often both recognize and reproduce the same song. Based on the above analysis, it is likely that there is one place for recognizing a song, and another for producing it.

---

[40] As the same spelling may have multiple meanings, multiple concept neurons are connected to the same spelling.

Order of Learning to Speak versus the order of Construction

When an infant is learning how to speak, their first attempt is babbling. They are mirroring the sounds they hear. The next stage is single words – "Mama," "Dada." Then comes whole sentences.

Below, the left side illustrates the order of learning to produce speech. The right side shows the order of processes in producing speech.

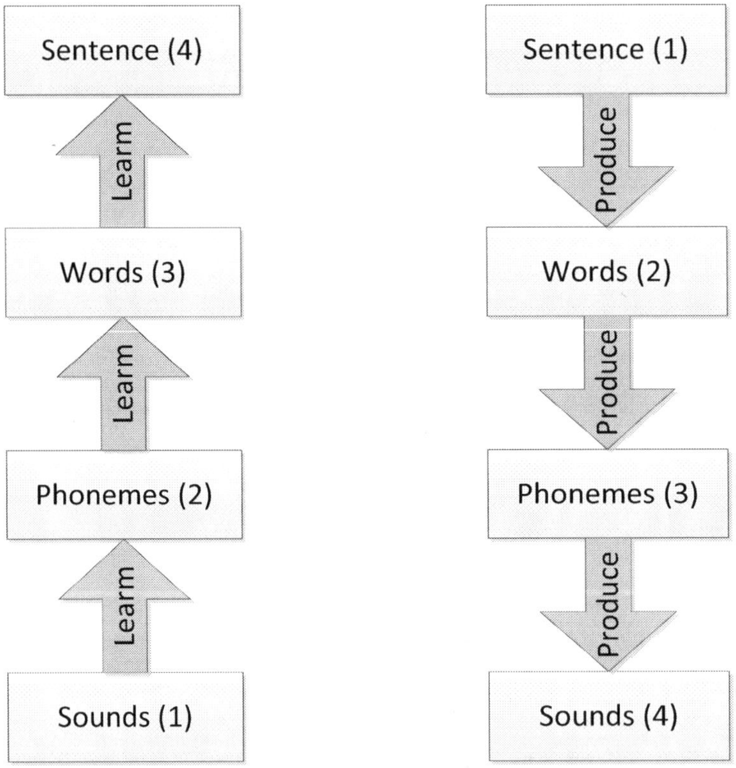

Words to sentences - Speech

A boy is running. The diagram below shows the translation from concepts into a sentence. The properties of both the boy and run concepts include a part of speech property - noun part (NP) or verb part (VP). Those properties determine the order of words in the sentence.[41] Gating neurons direct the output to be spoken, written, or typed.

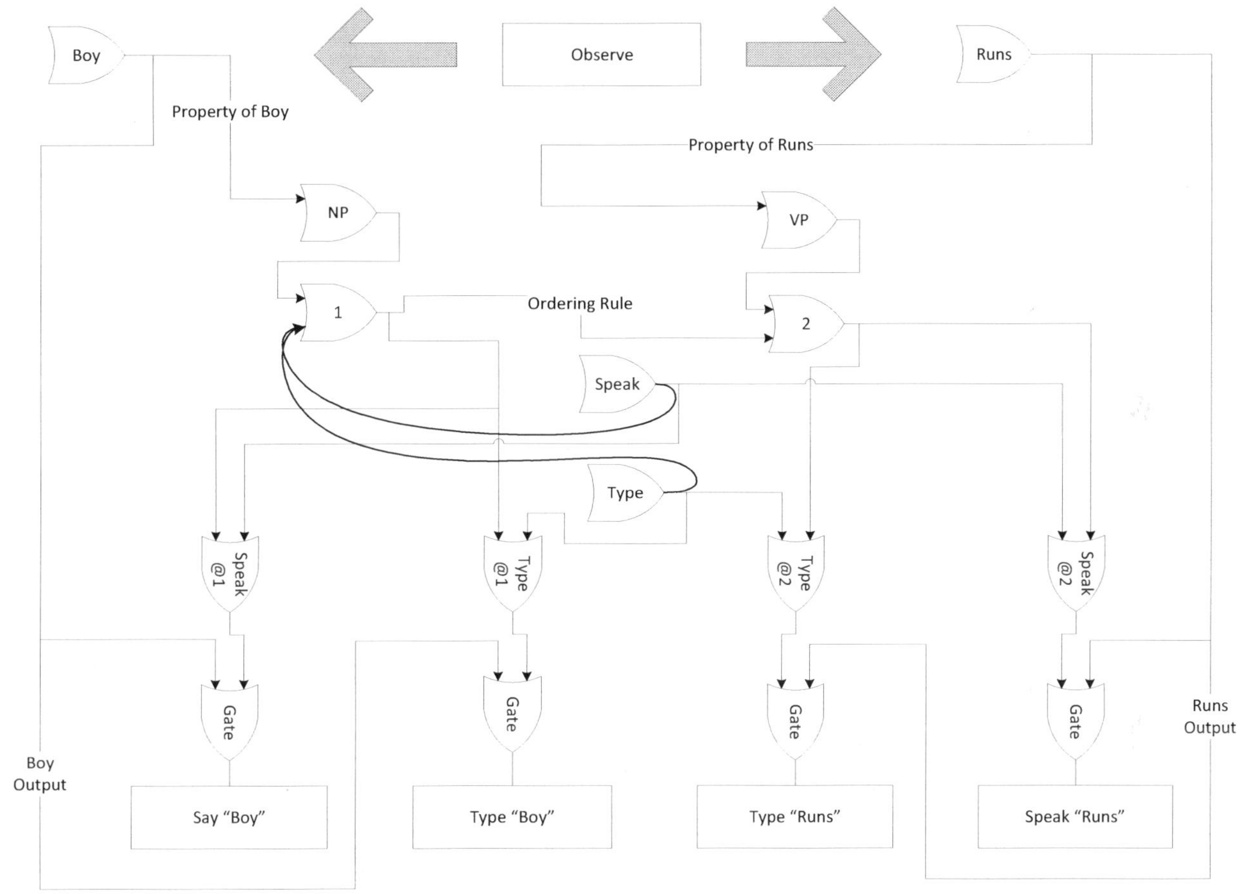

Typing a word means striking the right keys on a keyboard in the right order. Speaking a word involves uttering phonemes, described below.

---

[41] English is an SVO language, meaning the subject (boys) occurs before the verb (runs). Other languages include VSO, VOS and OVS – see https://en.wikipedia.org/wiki/Subject-verb-object. Also note that the above example and similar ones that follow are incomplete, in that a word such as boy could be part of the object component of a sentence.

Adding another word to the Sentence

Add the word Home to the sentence to make the sentence "Boy Runs Home." Then, the entire Verb Phrase gets processed as per below.

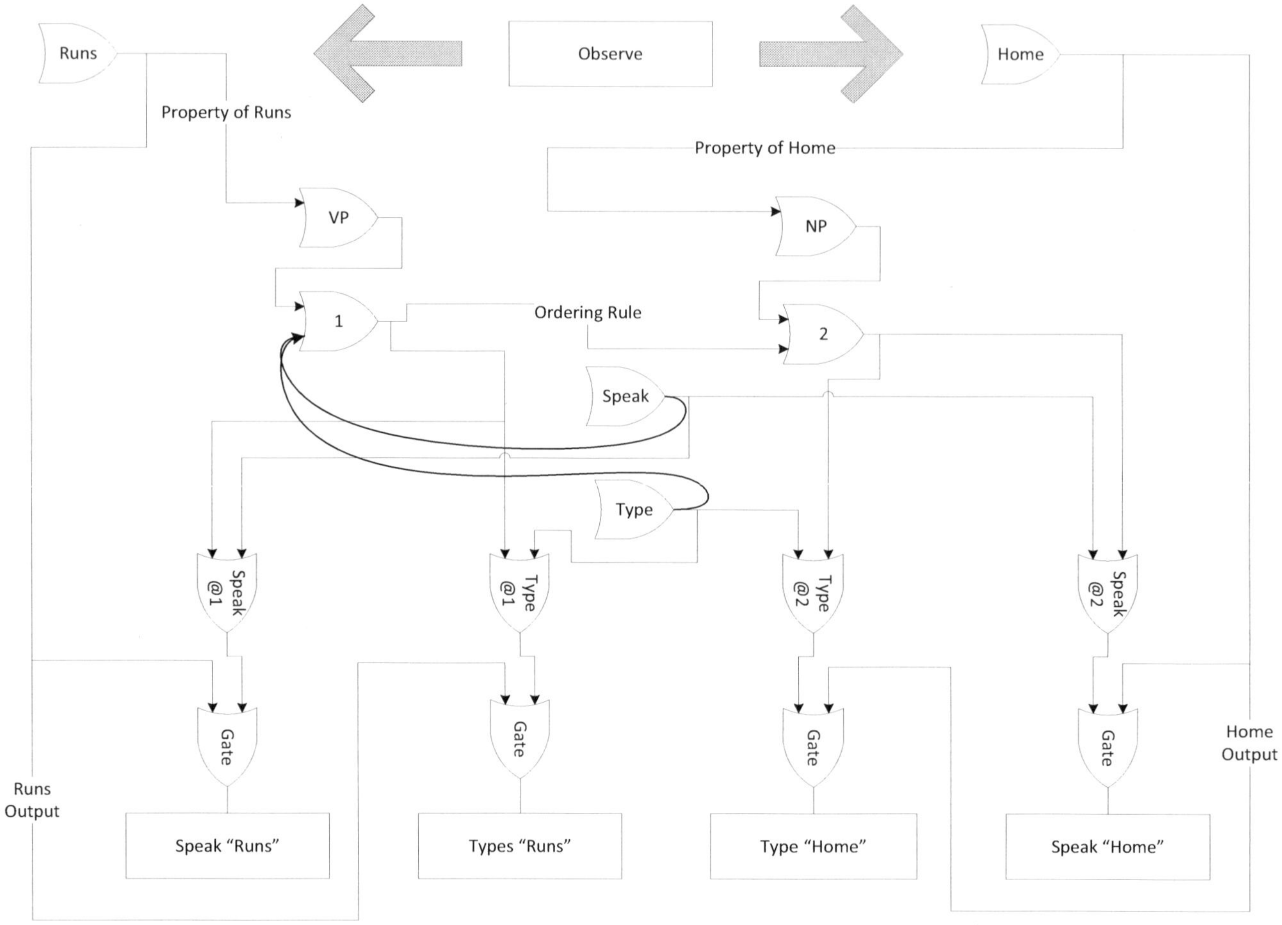

Word to Phonemes

The word concept neurons include a connection to the sequence of phonemes used to say the word. There are three phonemes in saying the word boy in a sequence connected to the boy concept neuron.

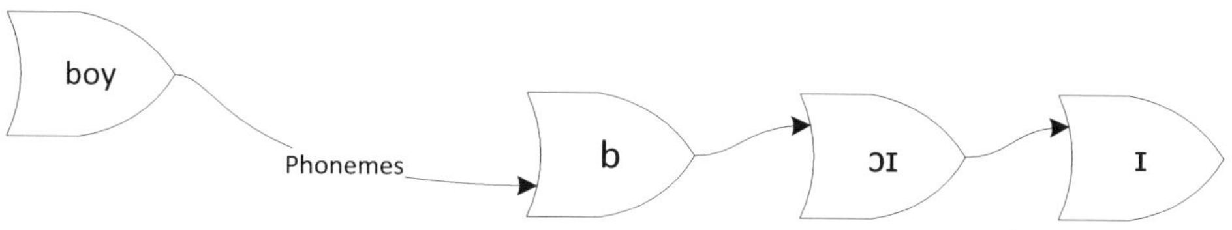

Phonemes to sounds

The phonemes are themselves concept neurons that have properties that direct motor neurons to produce sound[42]. Examples below:

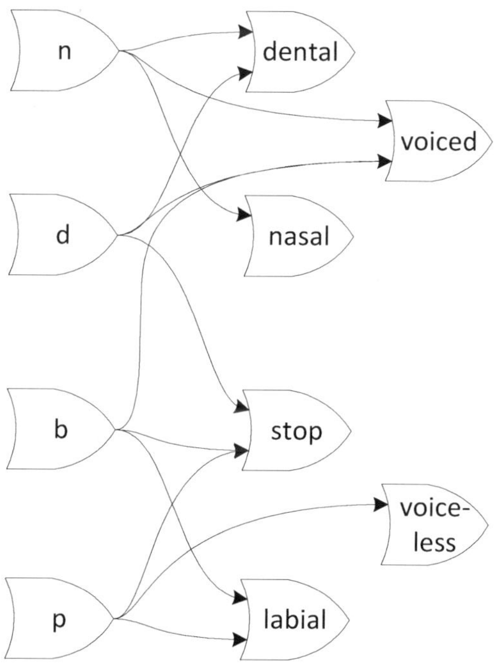

What the properties above mean:

<u>Labial</u>: one or both lips are the active articulators.
<u>Dental</u>: the tongue is against the upper teeth
<u>Stop</u>: vocal tract is blocked so that all airflow ceases
<u>Nasal</u>: allowing air to escape freely through the nose
<u>Voiced</u>: vocal cord is vibrating

[42] Phoneme properties are taken from page 25 of "Generative Phonology" by Sanford Schane.

Inner Monologue

There is another alternative path for the output of a sentence. It may be redirected back into another area of the brain – near where speech is analyzed. See the To-Self choices, "Boy" and "Runs" below.

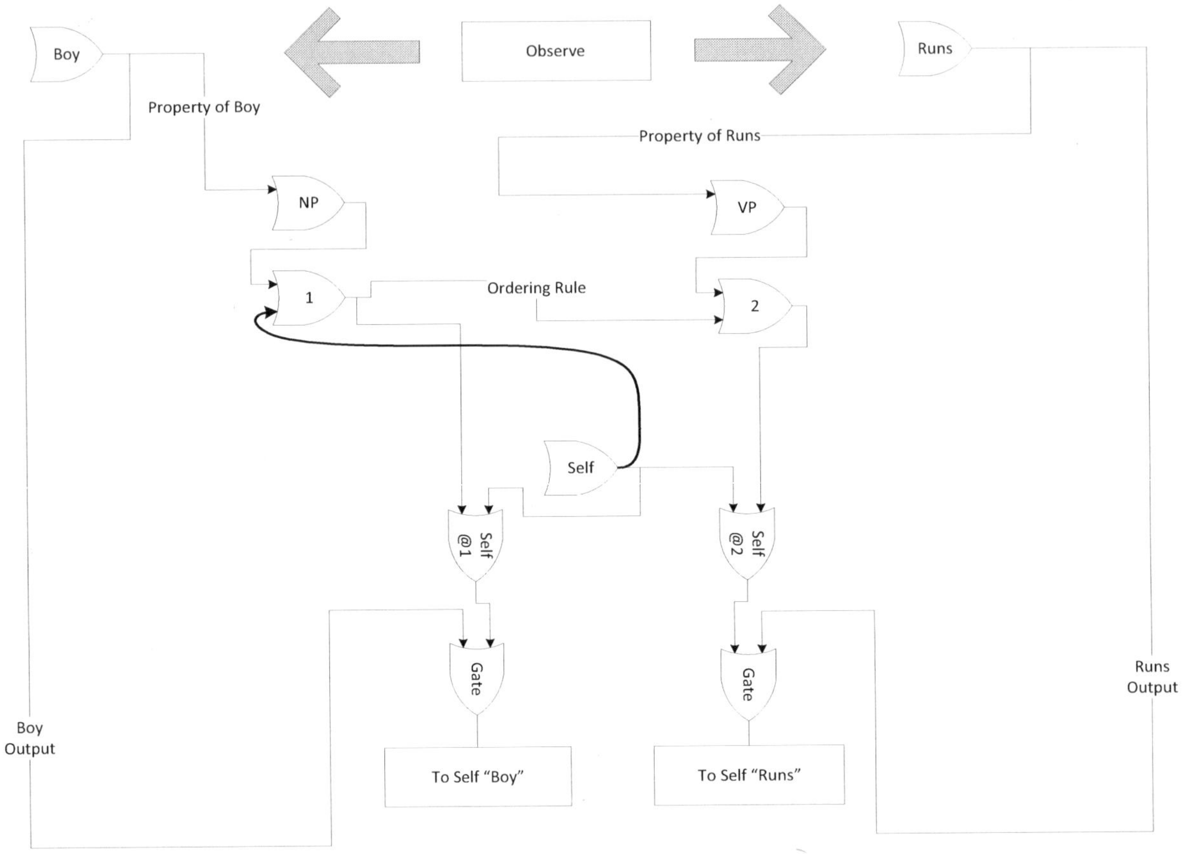

The Inner Monologue is that (nearly) continuous voice that narrates your life. In the diagram below, the output from the Create Ordered Sentence process is fed directly into the Analyze Sentence process. With only one analyze-sentence mechanism, you can't both listen to another person and yourself at the same time. Even planning what to say next may disrupt your listening to another person.

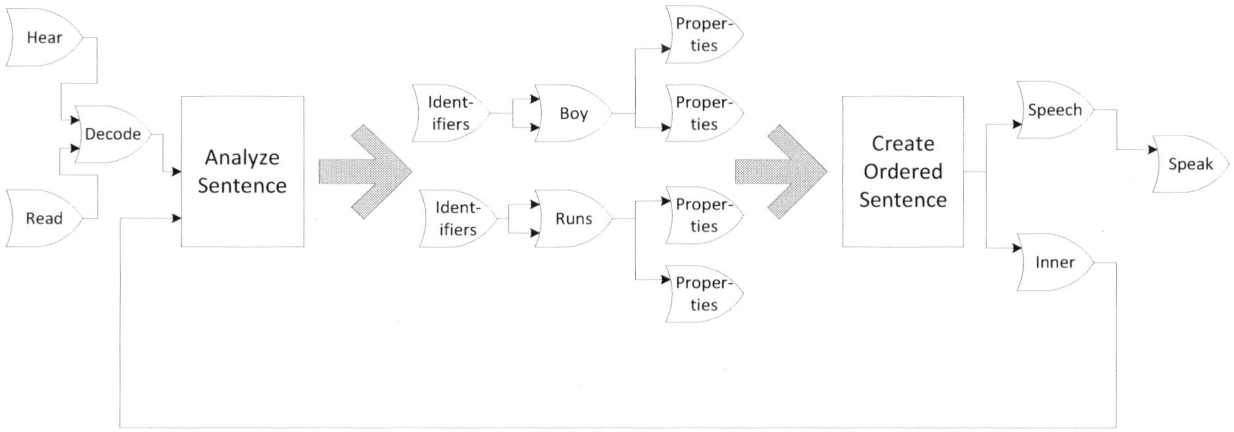

When you are stuck, you might ask yourself – "What Now?". Sometimes this helps. The reason may be that your ordinary attempt at resolving goals is not scouring the right places. The sentence analyzer above will direct the search to concept neurons, neurons that have words as identifiers.

I have noticed is that I find it much harder to write (or type) when I am listening to music with human voices. Perhaps writing involves, for me, the inner monologue, which shuts down the ability to hear the spoken word in music[43].

[43] Language is produced in Broca's area. Wernicke's area processes spoken words. The large bundle of nerves, the arcuate fasciculus, runs between the two. Perhaps the inner monologue uses this path?

Private Speech[44]

Another form of internal dialogue is "Private Speech."

At a young age, children sometimes talk to themselves[45]. They are speaking at a low volume, so only they can hear themselves. Perhaps:

1. They are practicing speaking with others.
2. Their ability to produce speech is more advanced than their ability to analyze the speech of others, and this helps them improve that other ability.
3. It develops before the inner monologue and confers some of the same advantages.

Intuition – Concepts without Words?

When my children were younger, I would watch Disney's Duck Tales with them. There was one episode where, at each turn of events, I knew what was going to happen next. I was certain that I had not watched this episode previously.

It was many years before I figured out the mystery. While downsizing, I encountered a Disney Duck Tales comic book. It had the same story as the TV episode. Mystery solved.

My recollection of the story was triggered by matching events. It was recognition knowledge, not explicit knowledge. Nor were words involved.

Many processes in life do not involve words. Dancing, climbing a rope, riding a bike do not involve words. Explaining these actions in words doesn't do them justice.

A goal can be satisfied by neurons that do not have any words attached. When you come up with an answer, but can't explain it in words, it is often called intuition.

Language – Good for what?

The goal of verbal language is to copy concept neurons and connections from one individual into another. While the mechanism is rough, the range of transmittable knowledge or behavior is much larger than mirroring. Written language has been key to our accumulation of knowledge.

Our ability to view and plan, across time, is the other element that has allowed humans to add together their contributions. We view time as if it is a physical construct. We place events on the calendar as if we were placing objects on a grid array. Both language, and the ability to coordinate efforts across time, are key advantages we have over other animals.

Next up are implications of the overall structure of the Neocortex.

---

[44] It has been pointed out that private speech is not the sole province of young children. Adults also engage in talking to themselves, at times. With the advent of mobile phones, it can be difficult do discern those having a conversation with others, and private speech.

Implications

Nature vs. Nurture

When examining a human brain, how can we be sure the differences are due to nature - differences at birth, or nurture – differences in the experience?

The brain of Albert Einstein has been studied, looking for anatomical differences. Studies have indicated that his regions involved in speech and language are smaller, while regions involved with numerical and spatial processing are larger[46]. Specifically, Einstein's lower parietal lobe (which is involved in mathematical thought, visuospatial cognition, and imagery of movement) was 15% larger than average.

The challenge with interpreting Einstein's brain is that we do not know whether Einstein's brain was different at birth, or whether it grew as he exercised that part of his brain.

Another example of a relationship of structure to function is the observation that a significant reduction of gray matter in part of Broca's area of those with Autism and Asperger's syndrome[47].

I know of a family where the parents, and the children, are all wonderful violin players. While the parents may have been attracted to each other by this common interest, how is it that all the children are so skilled? How can we disentangle nature from nurture?

Twin studies - studies of twins separated at birth, provide some insight. Anecdotally, there are suggestions that some twins have very similar preferences or careers, but not all. The likely conclusion is that both nature and nurture contribute, but neither factor is controlling.

Until we can monitor the detailed functioning of intact human brains from birth onward, disentangling nature from nurture will continue to dodge clear resolution.

Critical Periods

Consistent with the Hierarchical organization of the brain, development processes follow a logical progression, building new skills on existing ones. For example:

1. Recognize numbers -then- Addition and Subtraction -then- Multiplication and Division
2. Recognize letters -then- Words -then- Sentences
3. Musical notes -then- Musical Compositions -then- Styles of Music
4. Move limb-then- Move limbs smoothly in a sequence -then- Perform Ballet

The output of the more primitive skill feeds into the input of the more advanced ones. The output – the axon – must terminate closely – on the dendrites of the more advanced ones. More advanced skills are higher up the abstraction ladder.

In infant children, there are well established critical periods. Particular skills and abilities are acquired in sequence and according to a schedule.

---

46 https://en.wikipedia.org/wiki/Albert_Einstein%27s_brain

47 From page 50 of 'Cortical Functions' Reference of Trans Cranial Technologies.

The developmental sequence is consistent with our understanding of how the neocortex is laid out, with a progression of increasingly abstract functions. The later skills and abilities are of a higher level of abstraction and require that the earlier skills and abilities are firmly in place.

Education and experience are the sources of concepts for each level. If there is no exposure, or the lessons are not absorbed, one cannot build more abstract conceptual levels.

Learning a second language is much harder later on, as it is much harder to go back and adjust the connections that are tied to your first language, as you have built so many connections on top of them. Again, if you are not exposed to music early on, you will have more difficulty than most in being able to play a musical instrument or sing well.

That later abstractions build upon earlier ones mean that as we get older, the connections that we operate are a firmly established web of connections that reach down to our earliest levels. It is not just hard to handle change; it may be impossible to change the underlying assumptions and beliefs we have built up. Our worldview comes from what we experienced while forming those earlier connections.

Clever and Creative

The television show MacGyver features a very clever individual, Angus MacGyver. In light of the influence of the show on popular culture, the Oxford Dictionary now includes the definition:

> to "MacGyver" is to make or repair something" in an improvised or inventive way, making use of whatever items are at hand," e.g. "he MacGyvered a makeshift jack with a log."

The particular example in the definition points to the difference in connections in the neocortex. To MacGyver, a jack is not a specific mechanical device, but a lever used to lift a car, and a log has the rigid properties of a lever.

The properties and identifying connections of concept neurons are different in a clever person. Instead of specific 'things,' the connections are more abstract. In short, being clever is seeing the world through a more abstracted lens.

The more general notion, "creative," employs similar means. Creative individuals see things differently because the connections between concepts are at a higher level of abstraction. Creative individuals thus connect things that others do not, within the area of the neocortex, which is so constructed.

The conscious experience follows immediately below.

The Conscious Experience

Consciousness Defined

An enduring mystery is the nature of consciousness. The definition below largely confines consciousness to activity within the neocortex. Primary characteristics/requirements of conscious experience are:

1. A comparatively high level of neural activity, as measured by
   a. Electroencephalogram (EEG), Trans Cranial Magnetic Stimulations (TMS) measurements.
2. A clear boundary between self and non-self, i.e.
   a. Sensory inputs explicitly define and divide self from non-self. Sensory inputs from the skin, internal organs, and limb positions (proprioception) are self. Hearing and seeing yourself define and divide self from non-self.
3. A continuing experience of self
   - From moment to moment, a continuous sequence of events

What is not part of Consciousness - Sentience

Sentience is not consciousness. Sentience is the capacity to feel, perceive, or experience subjectively. Feeling incorporates emotions, which arise from below the neocortex. While the neocortex is structured uniformly throughout, most organs below the neocortex have distinct and different organizing principles[48]. Most likely, the particular organization and connections are the result of evolutionary pressures that began with the first central nervous systems.

As early as 1951, Wilder Penfield probed the surface of the somatosensory cortex. He established connections between locations on the somatosensory cortex and regions of the body. Recent research has established locations in the somatosensory cortex that correspond to emotional experiences[49], explaining expressions like a 'pit in my stomach,' or an aching heart. Hence, there are neural connections from the source of the emotions below the neocortex into the somatosensory cortex. Emotional experiences flavor the conscious experience.

What is not part of Consciousness – The Inner Monologue

The inner monologue narrates our existence. It is the voice-track of our lives. Yet, some conscious humans do not have verbal language. It is not essential to consciousness.

What is not part of Consciousness – Rapid Eye Movement (REM) sleep

Rapid Eye Movement or REM Sleep includes a high level of neural activity and continuing experience of self, but it cuts off all sensory input. Thus, it fails the above definition of consciousness. Without sensory input to stabilize the internal model of the external universe, the perceptual process goes haywire, often creating emotion-laden experiences.

---

[48] The cerebellum has a very uniform organizing structure.
[49] Bodily Maps of Emotions, PNAS January 14, 2014 and online at https://dx.doi.org/10.1073/pnas.1321664111

Levels of Brain activity during the experience of consciousness

So, what is occurring in the brain during Consciousness? Old fashion computers had a single processing core and executed one process at a time. Current day computers execute multiple processes at the same time, on multiple processing cores. The brain is like modern computers in that it is executing multiple processes at the same time.

There are four independent processes identified below, all operating at the same time. The last one has four levels. Only one of those levels is active at any one time. Metabolic (caloric) requirements increase with level, and so the brain defaults to operating at the lowest level required.

The **Perceptual** process runs continuously while conscious - taking sensory input and producing an internal model of external reality—expectations of what is to come next direct analysis of sensory input.

Concurrent with

A comprehensive **prediction** of the future state of the universe, incorporating what you observe and what you are going to do. Errors in prediction result in new concepts and connections that reflect the new information of how the universe works.

Concurrent with

In mature adults, a region of the Prefrontal cortex monitors intended actions. When it identifies an action that would lead to a poor outcome, such as jumping off a cliff or telling your boss off, it **suppresses** motor neuron activity. Maturation of this function in the prefrontal cortex generally occurs in the mid-twenties, hence the sometimes-poor decisions of teenagers.

Concurrent with one of

[**Reflection**] At **Level-1**, you are awake and conscious, but resting in place, with little attention to external events. You might be recalling past events, or planning for the future.

Or

[**Interacting**] At **Level-2,** you are actively interacting with the (external) universe. Walking, driving a car, or cooking a meal are examples. Automatic behaviors are running the show as long as the pre-conditions for their execution match current conditions. Because of the lack of thought needed for Level-2, you might also be recalling past events, or planning for the future, which is Level-1 activity.

Or

[Instruction] Level-3 occurs when you are actively engaged in learning. New information results in new concepts and connections, or the voiding of existing ones. Hopefully, you are having fun.

Or

[Alarm] Level-4 When the universe has presented you with conditions for which no automatic behavior exists, you need to figure out how to react. When you try something new, and it works, you add it to your collection of Automatic Behaviors. If not, you keep trying.

Conclusion

The limitations of our cognitive abilities derive from the two mechanisms that strengthen synapses – coincidence and prediction. Our ability to make use of the connections in behaviorally relevant periods arises from the ability of a neuron to retain focus.

These mechanisms have served us well as hunter-gatherers, and in our current, technologically advanced civilization. Is it possible that we need other schemes for conceptualization to understand the universe more completely?

The Connections Model ties together cognitive processes and integrates them into the experience of consciousness. The model fills the gap between the understanding of individual neurons and the high-level functions of the neocortex, up to and including consciousness.

The Connections Model makes only passing reference to structures below the Neocortex. Nonetheless, these underlying components of the Human brain are important in explaining human behavior.

There are other, competing models for the neocortex. In particular, success in predicting how the neurons that recognize faces (Face Patch Neurons) have strengthened an alternative approach, the Parallel Distributed Processing model. In the Addendum that follows, I describe the Face Patch Neuron model, and an alternative interpretation wherein the results from that support the Connections Model.

With so many neurons, the difficulty of locating the exact neuron that implements a concept is exceptionally difficult. Perhaps computer simulations of the Connections Model to provide insights not readily obtained by observing individual neurons in action. Moreover, the Connections Model paves the way for the creation of a generalized machine ~~artificial~~ real intelligence.

Addendum

Concept Neuron vs. Distributed Representation

In the now popular distributed representation model, concepts are the product of activation patterns of multiple neurons.

The Connections Model takes an alternate view that there is a one-to-one correspondence of a single concept to a single neuron.

One of the objections to the "concept" neuron is its biological implausibility. One reason is the belief that a single neuron cannot generate enough output to accomplish this. 'Of Mice and Men and Chandeliers' disputes this[50], wherein a single neuron's firing results in a polysynaptic chain of activity.

A broad examination of the status of this area of research appears in a paper by Jeffrey S. Bowers[51]. This paper points out the ambiguity in definitions that plague this area.

The next section describes the work of Dr. Tsao and her team. They researched the 'Face Patch Neurons.' Their result is a mathematical model that predicts the output of the Face Patch neurons. She has taken that result to mean that the single neuron concept model is invalid, as no single neuron in the output corresponds to an individual's face.

After describing her work, I point to an alternate understanding of what Dr. Tsao found, and how that is consistent with the wiring for recognizing faces presented earlier. Her conclusion is premature.

---

[50] Woodruff A, Yuste R (2008) Of Mice and Men, and Chandeliers. PLoS Biol 6(9): e243. https://doi.org/10.1371/journal.pbio.0060243
51 Jeffrey S. Bowers (2017) Grandmother cells and localist representations: a review of current thinking, Language, Cognition and Neuroscience, 32:3, 257-273, DOI: 10.1080/23273798.2016.1267782. https://doi.org/10.1080/23273798.2016.1267782

Face Patch Neurons (FPN)

Face Patch neurons are integral to the recognition of faces. Face Patch neurons are a string of blueberry-size regions on each side of the brain. A team, led by Dr. Doris T. Tsao first located these structures and inserted electrical probes to detect the firing – output – of 205 Response Face Cells/neurons at the end of the Face Patch neuron (FPN) chain.

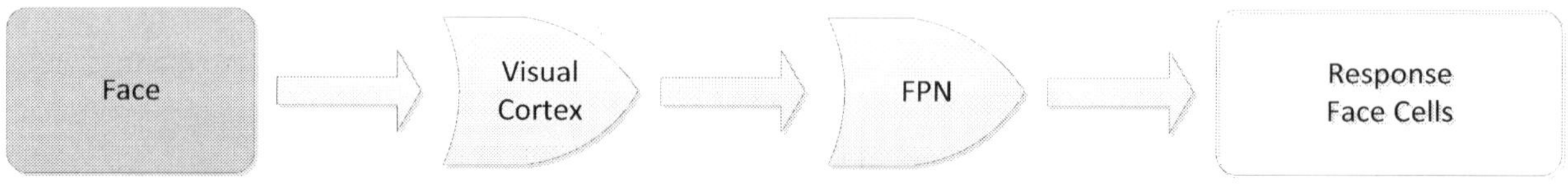

The team wanted to understand the relationship between the input and the output in the diagram above. The output was already in numerical form. A numerical representation of the facial characteristics was needed.

The team created a methodology based on the 'Active Appearance' approach used by artificial intelligence programs that recognize faces. Like the Active Appearance approach, they also broke down facial features into two broad categories - shape and appearance. Properties that derive from the skeletal are examples of shape features, such as the width of the forehead and the spacing of the eyes. Appearance features include complexion, eye, hair, and skin color.

Through a multistep process, they derived 25 shape feature values and 25 appearance feature values that uniquely and precisely characterize facial images. From those parameters, they created faces of every shape, size, and character to present to a subject's visual system, as illustrated below:

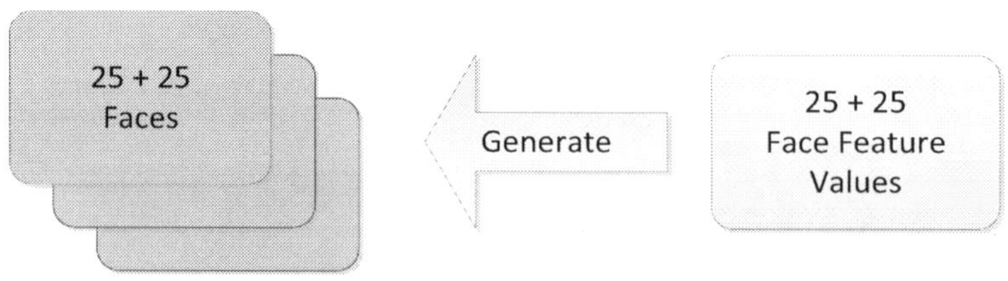

The subject viewed the generated facial images while recording the 'Response Face cell' voltage levels.

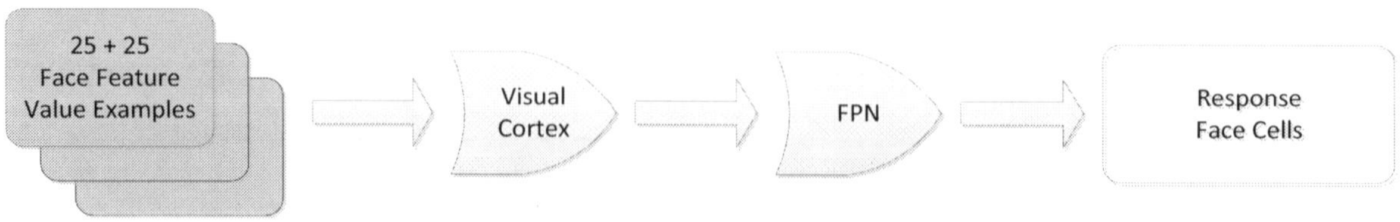

Reviewing the data showed a simple relationship between the Response of the Face Cells and the features of the face viewed. The inverse relationship, as illustrated below as well, was the mathematical inverse.

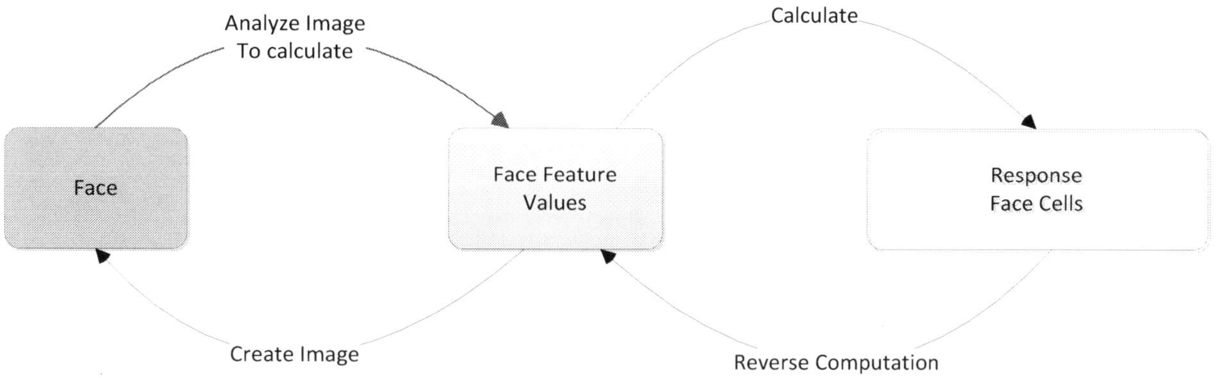

While a computer program can identify the "Face Feature" values from any face, or create a facial image from "Face Feature" values, the right half is simple matrix multiplication, i.e.

$$Response.Face.Cells = Weight.Matrix * Face.Feature.Value$$

$$Face.Feature.Value = (Inverse\ of\ Weight.Matrix) * Response.Face.Cells$$

While the results are impressive, it is clear that the underlying mechanism of the actual Face Patch neurons is not a series of computations. Neurons do not add and multiply. Yes, the calculations produce the same result, but that does not clarify how the Face Patch neurons perform the recognition process.

Even so, the success of this model has led some to believe that these results disprove the notion of grandmother cells, closely related to the concept neurons of the Connections Model. This conclusion is premature.

Instead, the results confirm the ideas presented in the Connections Model for the recognition of faces, where the 205 Response Face Cells correspond to the mid-level neurons, diagrammed in the chapter "Perception – Recognizing Objects (Ventral Stream)" section: "Recognizing Combinations of Features – Faces."

Deep Learning

A current popular area in Artificial Intelligence is Deep Learning. Biological neural networks are the inspiration for computational neural networks. Unlike biological neural networks, computational neural networks perform addition and multiplication.

In a simple computational neural network, there is an input layer, an output layer, and one or more hidden layers in between, as shown below.

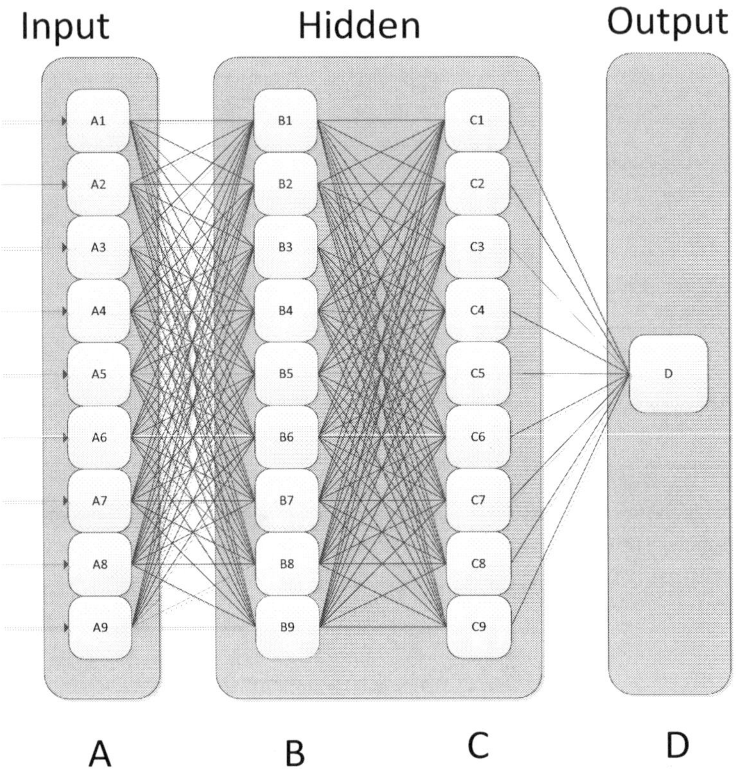

The process is as follows:

1. The 'A' layer receives input data (numbers); and then
2. The 'B' layer values are determined using the 'A' layer values; and then
3. The 'C' layer values are determined using the 'B' layer values; and then
4. The 'D' layer values are determined using the 'C' layer values.

Below is the calculation of the B4 element.

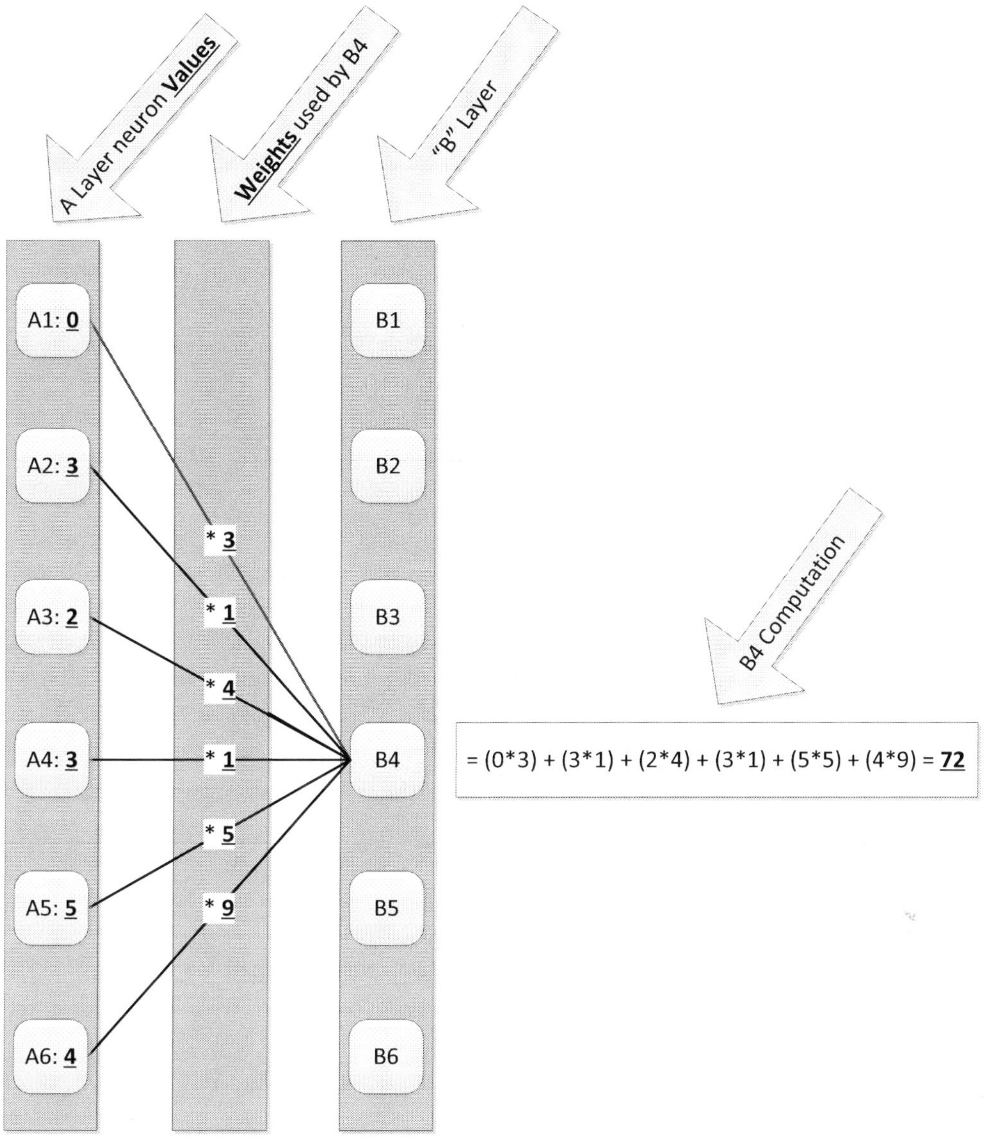

For the Mathematically inclined, the calculation for the B level neurons:

$$B_j = \sum_{i=1}^{n} A_i * Weight_{ij}$$

Note that the more general calculation for an arbitrary sized computational neural network is matrix multiplication.

Google uses computational neural networks to translate between languages and in picture processing. Google created Google Brain – special hardware that performs the calculations of neural networks. Google Translate and Google's Image recognition utilize Google Brain.

The first step in creating a useful neural network is to train it. Training involves feeding lots of sample data into the input side and modifying the weights to match the desired results. When training the neural network below, the weights into the 'D' layer would be adjusted first, then the weights into the 'C' layer, etc.

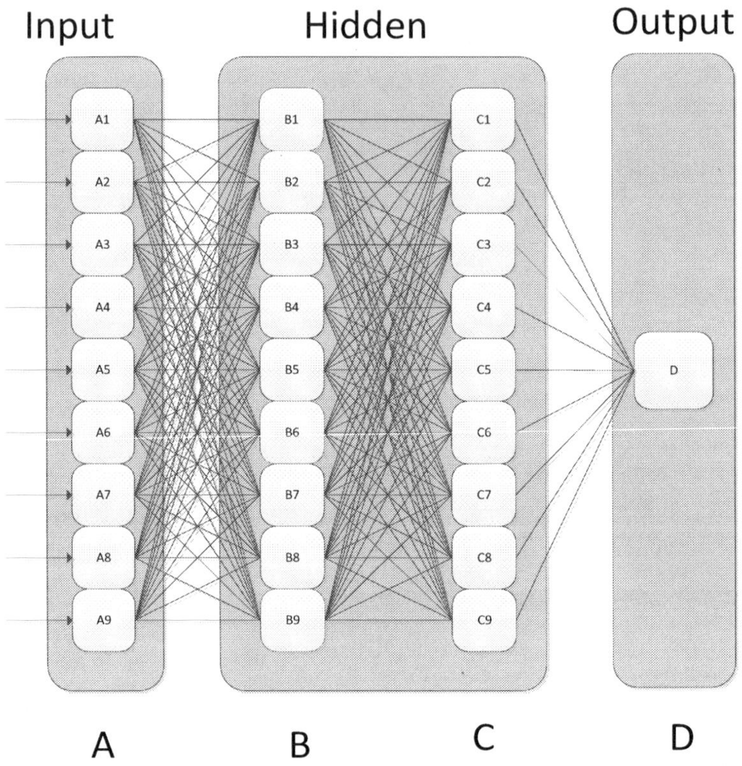

We can train the neural network to recognize dogs. When there is a dog in a picture, the 'D' neuron will light up.

One challenge in dealing with computational neural networks is that we do not understand how they produce their results. There is no clear relationship between the in-between neurons, their weights, and the task they are performing.

In that they are being considered as part of the mechanism for self-driving vehicles, we will need to develop strong evidence that they are not confused by unexpected inputs.

About the Author

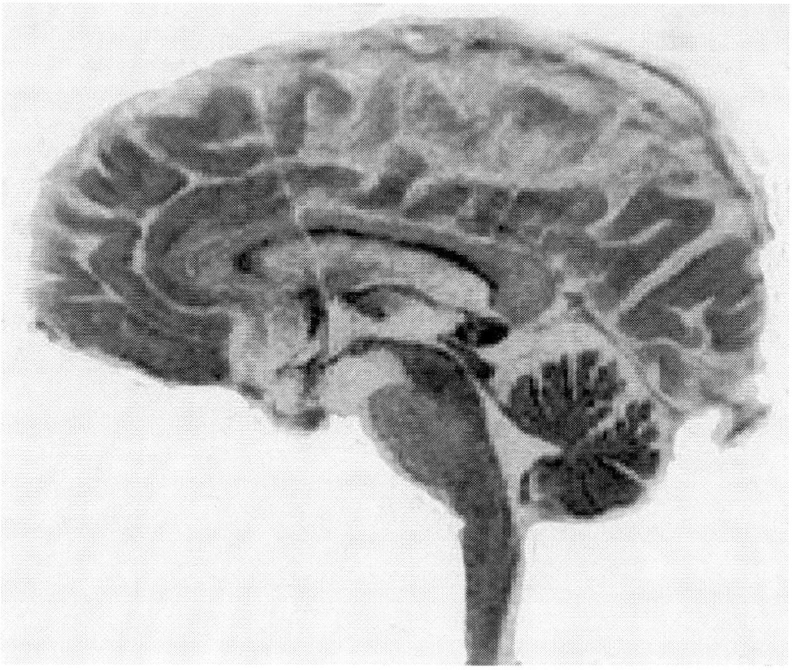

Bruce Maier wrote his first computer program at age 14 in 1968. In 1971 he created a compiler for a new programming language, Cryptk, which ran on the Long Island Regional Instructional Computer System, or LIRICS. In 1976 he received a Bachelor in Computer Science from Stony Brook University. During the 1970s and early 1980s, his firm created a Database Management System and implemented integrated applications. In 1988 he joined the New York City Transit Authority and led application development projects. After retirement, he applied his information system design skills in constructing the Connections Model, the subject of this book.

Bruce lives in Shoreham, New York, with his wife, Sandra. He has three children, James, Gregory and Samantha, and three grandchildren, Noah, Timothy, and Joshua.

The sagittal view of the brain of the author is pictured above.

Coming soon to Amazon

Reality is Virtual, 2nd Edition

By Bruce Maier

When we understand all of the physical laws of the universe, will we be able to create a computer simulation that is indistinguishable from reality? Digital computers have limitations. Do those limitations preclude a perfect simulation, or do they explain?

Why is it impossible to go faster than the speed of light?

Why does light behave like it is a wave at times, and like a particle at other times?

Why are there three spatial dimensions?

What if the universe is a computer simulation? And if it is, what is outside of it?

*First Edition cover is depicted above.

Made in United States
Troutdale, OR
11/03/2026

56844523R00073